Understanding Proof

T0206606

$$P(A|B) = \frac{P(A \cap B)}{P(B)}$$

(x)

$$2ab \leq a^2 + b^2$$

$$+ \frac{f(x_2)}{2} + \ldots +$$

$$\frac{d\sin(x)}{dx} = \cos(x)$$

$$\sin^2\left(\right.$$

Tarquin

Figure 4.1(a), Page 23, Public domain;
Figure 4.1(b), Page 23, Public domain;
Figure 4.2, Page 31, Shutterstock Image;
Figure 5.1, Page 47, Public domain;
Figure 6.1(a), Page 65, Public domain;
Figure 6.1(b), Page 65, Public domain;
Figure 7.1, Page 96, Public domain;

©2018 Tarquin Publications and the Authors
Book: ISBN
EBook: ISBN

Printed by LSI Globally
All rights reserved.
Available in the USA www.ipgbook.com

Tarquin Publications
Suite 74, 17 Holywell Hill
St Albans
AL1 1DT
UK
www.tarquingroup.com

Contents

1. Introduction to Proof

Throughout our early schooling we are taught that mathematics is about using and manipulating numbers, perhaps with some application in mind and sometimes with seemingly none. Of course, this is a part of what mathematics is about, but it is not the whole story. A professional mathematician is not somebody who sits around thinking about large numbers, for example. A mathematician is somebody who formulates *conjectures* about observations and then tries to show if these conjectures are true or false. In this way, a mathematician is similar to any other scientist, but rather than performing experiments to *prove* their claims, mathematicians rely on logic and reason in their *proofs*. This is perhaps the distinction between mathematics and other sciences, once a statement has been *proved mathematically*, the logic is undeniable and the statement will remain true forever more; the conjecture has become a *theorem*.

Perhaps the most familiar theorem of all is Pythagoras' theorem that relates the lengths of the sides of a triangle. Pythagoras lived in the 6th century BC (although it is likely his theorem was known well before this), but his theorem remains as true today as it did all that time ago and it is just as useful, not just for its applications, but also for providing the building blocks for the proof of more elaborate conjectures and theorems. In contrast, other scientists build up a weight of evidence in support of a *theory*, but the theory is never proven for definite. A classic example is Newton's Laws, which provide a good approximation the movement and interaction of objects under certain circumstances, but they were improved and extended by Leonhard Euler (*ca.* 1750) and then again by Albert Einstein in the early 20th century when he developed his theory of Special Relativity. In other sciences, theories evolves, whereas, in maths they remain for all time.

It might be tempting to think that, with the history of mathematics stretching back many millennia, nearly everything has already been proved. This is actually far from the truth; as time has gone on, more and more branches of mathematics have developed and more and more questions have been posed that remain unanswered. In the year 2000, the Clay

Mathematics Institute offered prizes of $1 000 000$ for anybody who could solve one of 7 outstanding mathematical problems (https://www.claymath.org). To this day (February 2018), only one of these problem has been solved, the Poincaré conjecture. The large sums of money involved show how important these mathematical questions are and that it is possible to earn a lot of money from mathematics.

This book is aimed at mathematics students who are just becoming acquainted with the concept of proof and the *rigour* required when proving something. Clearly, there are very few results that an early student can hope to prove, but being exposed to the proofs behind well known results will help a student gain a deeper insight and understanding of those results and develop their capacity for logical thought. For some students who go on to study mathematics at a higher level, this knowledge of how to formulate a proof will be essential and exposure to the different methods of proof will be invaluable.

Each area of mathematics begins with a few results, which are either undeniable, impossible to prove or just act as a starting point. These are called *axioms*; for example, the *Peano axioms* define the arithmetic properties of natural numbers and include statements such as '0 is a natural number' and 'if x and y are natural numbers, then $x = y$ implies $y = x$'. Once we have a set of axioms, we can start to define other interesting objects and then begin to prove new theorems about these new objects, which might in turn lead to more complex definitions and further proof. A rich tapestry of mathematical ideas can quickly be built in such a way. In order to prove new results, it is essential to have a deep understanding of what has gone before and how proofs were set out. This is why studying simpler proofs is essential for any mathematician. In some instances a proof can be formulaic and a systematic approach will work, but, more often than not, proofs require some creativity.

In this book we introduce the common techniques of mathematical proof, including *direct proof*, *proof by contradiction* and *proof by induction*. We shall also explain how these methods can be applied to other areas of mathematics in the syllabus, for example, to geometry and trigonometry, calculus and statistics. In doing so, we also touch on proofs that provide an excellent grounding for the student who intends to further their mathematical or scientific study.

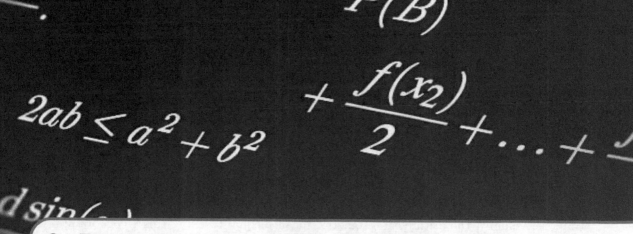

2. Exploring Methods of Proof

As an introduction to methods of proof, we consider a number of different ways of approaching the proof of the arithmetic-geometric mean inequality. The methods presented in this chapter will be explored in more depth throughout the rest of the book.

Before we can state the arithmetic-geometric mean inequality, we must define what the *arithmetic mean* and *geometric* means are.

Definition 2.1 — Arithmetic Mean

Consider two real numbers a and b, the arithmetic mean of these numbers is

$$m_a = \frac{a+b}{2}.$$

Notice that the numbers a, m_a and b form an arithmetic sequence.

Definition 2.2 — Geometric Mean

Consider two real numbers $a \geq 0$ and $b \geq$, the geometric mean of these numbers is

$$m_g = \sqrt{ab}.$$

In this case the name arises due to the numbers a, m_g and b forming a geometric sequence.

We now state the theorem.

> **Theorem 2.3 — The Arithmetic-Geometric Mean Inequality**
> For two real numbers $a \geq 0$ and $b \geq 0$, the arithmetic-geometric mean inequality is:
>
> $$\frac{1}{2}(a + b) \geq \sqrt{ab}. \tag{2.1}$$

> **Interactive Activity 2.1 — The Arithmetic-Mean Inequality**
> Explore, visually, the arithmetic-geometric mean inequality using this Geogebra app.

2.1 Direct Proof

In this direct proof, we start with known facts and perform a number of logical steps/deduction until the we reach the desired result. Our starting point will be the fact that $(a - b)^2 \geq 0$, as any squared number is non-negative.

$$
\begin{aligned}
(a - b)^2 &\geq 0, \\
\Rightarrow \quad a^2 - 2ab + b^2 &\geq 0, \\
\Rightarrow \quad a^2 + 2ab + b^2 &\geq 4ab, \\
\Rightarrow \quad (a + b)^2 &\geq 4ab, \\
\Rightarrow \quad \frac{1}{4}(a + b)^2 &\geq ab, \\
\Rightarrow \quad \frac{1}{2}(a + b) &\geq \sqrt{ab}.
\end{aligned}
$$

> **Proof Tip**
> It is common to see "proofs" similar to the following.
>
> $$\frac{1}{2}(a + b) \geq \sqrt{ab}$$
> $$\frac{1}{4}(a + b)^2 \geq ab$$
> $$(a + b)^2 \geq 4ab$$
> $$a^2 + 2ab + b^2 \geq 4ab$$
> $$a^2 - 2ab + b^2 \geq 0$$
> $$(a - b)^2 \geq 0$$
>
> Here the result has been assumed and then a sequence of deductions to a true statement has been made. This is not valid; to prove something in a deductive fashion we must proceed to the result, not start from it. We must also provide some indication of how we have gone from one line to the next, which has not been done in this case.

2.2 Graphical Proof

The figure below can be interpreted as a proof of (2.1). However, this requires some understanding on behalf of the reader and it is preferable to write some words as guidance.

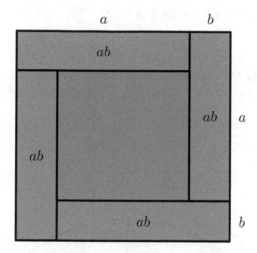

The larger square (of side length $a + b$) has an area greater than the sum of the area of the four rectangles for any $a \neq b$. If $a = b$ then the areas are the same. Hence,

$$(a + b)^2 \geq 4ab \qquad \Rightarrow \qquad \frac{1}{2}(a + b) \geq \sqrt{ab},$$

proving the result.

The triangle shown below can be used to provide an alternative proof of (2.1) using geometry.

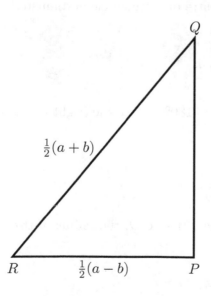

Since the triangle RPQ is right angled, we can apply Pythagoras' Theorem to find the length of the side PQ.

$$\begin{aligned} |PQ| &= \sqrt{\left(\frac{1}{2}(a+b)\right)^2 - \left(\frac{1}{2}(a-b)\right)^2} \\ &= \sqrt{\frac{1}{4}(a^2 + 2ab + b^2 - a^2 + 2ab - b^2)} \\ &= \sqrt{ab}. \end{aligned}$$

Since the hypotenuse RQ will always have a length greater than that of the perpendicular PQ (except in the degenerate case when they coincide) we have that $\frac{1}{2}(a+b) \geq \sqrt{ab}$, thus proving the result.

Finally, we can also use the image shown in the Geogebra interactive above to derive a proof.

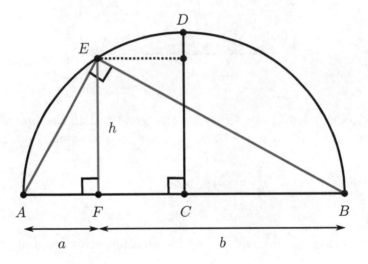

In the image above, C is the centre of a semicircle of diameter $a + b$. Hence,

$$|AC| = \frac{a+b}{2},$$

$$|BC| = \frac{a+b}{2}, |DC| = \frac{a+b}{2}.$$

Angles $\angle FAE$ and $\angle FBE$ sum to $180°$, and so the (right angled) triangles FEA and BEF are similar. Hence,

$$\frac{h}{a} = \frac{b}{h},$$

$$\Rightarrow \quad h^2 = ab, \Rightarrow \quad h = \sqrt{ab}.$$

Now, the length of h must be less than CD, the radius of the circle. Hence, the result is shown,

$$\sqrt{ab} \leq \frac{1}{2}(a+b).$$

Note that equality occurs when $a = b$.

> **Proof Tip**
> It could be argued that an advantage of this proof over others in this chapter is how easy it is to see that we have equality if $a = b$.

2.3 Proof by Contradiction

In proof by contradiction, we assume that the result we are trying to prove does not hold, then show that we arrive at a contradiction. Let us assume that the arithmetic-geometric mean inequality does not hold, *i.e.* $\frac{1}{2}(a+b) < \sqrt{ab}$. We follow the steps below to achieve a contradiction

$$\frac{1}{2}(a+b) < \sqrt{ab},$$
$$\Rightarrow \quad \frac{1}{4}(a+b)^2 < ab,$$
$$\Rightarrow \quad (a+b)^2 < 4ab,$$
$$\Rightarrow \quad a^2 + 2ab + b^2 < 4ab,$$
$$\Rightarrow \quad a^2 - 2ab + b^2 < 0,$$
$$\Rightarrow \quad (a-b)^2 < 0.$$

Since the square of any number is non-negative, the final line above is a contradiction and we can conclude that the result holds.

2.4 The Generalised Arithmetic-Geometric Mean Inequality

We can generalise the arithmetic-geometric mean inequality to the case where we are finding the mean of two or more numbers.

> **Theorem 2.4 — The Generalised Arithmetic-Geometric Mean Inequality**
> For n real non-negative numbers, $a_1, a_2, \cdots, a_n$, the arithmetic-geometric mean inequality is:
> $$\frac{1}{n}(a_1 + a_2 + \cdots + a_n) \geq \sqrt[n]{a_1 a_2 \cdots a_n}. \qquad (2.2)$$

This theorem can be proved using a technique known as mathematical induction. For this result we proceed as follows.

Let $P(n)$ be the statement "$\frac{1}{n}(a_1 + a_2 + \cdots + a_n) \geq \sqrt[n]{a_1 a_2 \cdots a_n}$"

Step 1: We prove the result for the base case $n = 2$. In this case the result is the same as the previous result we have proved, and so we know it to be true.

Step 2: We assume the result to be true for $n = k - 1$, $\frac{1}{k-1}(a_1 + a_2 + \cdots + a_{k-1}) \geq \sqrt[k-1]{a_1 a_2 \cdots a_{k-1}}$.

Step 3: We now show that the truth of $P(k)$ follows from the truth of $P(k-1)$.
Without loss of generality we assume that $a_1 \leq a_2 \leq \cdots \leq a_k$. For clarity we will denote

the geometric mean by GM, i.e. $GM := \sqrt[k]{a_1 a_2 \cdots a_k}$. It is clear that $a_1 \leq GM \leq a_k$, and since $a_1 + a_k \geq \frac{a_1 a_k}{GM} + GM$,

$$
\begin{aligned}
0 \leq a_1 + a_k - GM &- \frac{a_1 a_k}{GM} \\
&= \frac{a_1}{GM}(GM - a_k) + (a_k - GM) \\
&= \frac{1}{GM}(GM - a_1)(a_k - GM).
\end{aligned}
$$

By our inductive hypothesis,

$$
\frac{a_2 + \cdots + a_k + \frac{a_1 a_k}{GM}}{k-1} \geq \sqrt[k-1]{GM^{k-1}}
$$
$$
= GM.
$$

Hence,

$$
a_2 + \cdots + a_{k-1} + \frac{a_1 a_k}{GM} \geq (k-1)GM
$$
$$
\Rightarrow \quad \frac{a_2 + \cdots + a_{k-1} + \frac{a_1 a_k}{GM} + GM}{k} \geq GM
$$
$$
\Rightarrow \quad \frac{a_1 + a_2 + \cdots a_k}{k} \geq GM.
$$

In the last line above we used our established fact that $a_1 + a_k \geq \frac{a_1 a_k}{GM} + GM$. We have now shown that $P(k)$ follows from $P(k-1)$.

Step 4: Since we have shown the result to be true for $n = 2$, and that if the result is true for $n = k - 1$ it is also true for $n = k$, the result is true for all positive integers $n \geq 2$ by the principle of mathematical induction.

> **Remark**
> This is the simplest inductive proof of the generalised arithmetic-geometric men inequality that the authors are aware of. It is due to an article "The Arithmetic Mean?Geometric Mean Inequality: A New Proof" by Kong-Ming Chong pulished in 1976 in "Mathematics Magazine"

At first sight this proof appears complex; mathematical induction will be covered in more detail in Chapter 6.

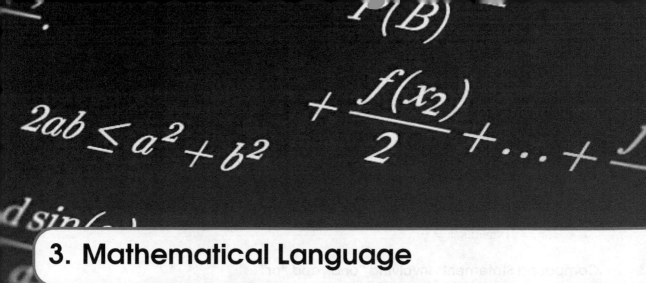

3. Mathematical Language

Mathematical arguments are precise and unambiguous. We must learn how to use mathematical language correctly in order to produce sensible mathematical arguments. In this chapter, we introduce and practise using mathematical language correctly.

.1 Statements and Predicates

We construct mathematical arguments using mathematical objects called *statements*, which can be either true or false, but not both. For example, the statement "$4 > 3$" is a true statement, whereas "$4 < 3$" is false. To make our lives easier, we often denote such statements with letters such as A, B, c, *etc.*

Many mathematical expressions contain variables so that, until we know the values of the variables, it is impossible to say whether the expression is true or false. In this case, the expression is called a *predicate*. The verity of the expression is *predicated* on the value of the variables. For example, $x > 0$ is a predicate; until we know the value of x, we cannot say whether the expression is true or false. If $x = 2$, then it is true that $x > 0$ and then $x > 0$ is a statement. Predicates are also often denoted by letters and the unknown variable, for example, $p(x)$, $q(y)$, where x and y are the variables.

.1 Negation

As we require statements to be true or false, we can introduce the notion of negation. Given a statement A, $\neg A$ is read "not A". Some examples are:

- If A: "It is raining" then $\neg A$: "It is NOT raining".
- If B: $4 > 3$ then $\neg B$: $4 \leq 3$.

Table 3.1 is called the *truth table* for negation, it shows that $\neg A$ is true (T) or false (F), when A is false or true, respectively.

A	$\neg A$
T	F
F	T

Table 3.1: Truth table for negation, $\neg$.

Some examples of negations of statements where x and y are (known) real numbers are:

- $\neg(x < y)$ means $x \geq y$,
- $\neg(x > y)$ means $x \leq y$,
- $\neg(x = y)$ means $x \neq y$.

3.1.2 Compound Statements Involving "and" and "or"

Two (or more) statements/predicates can be combined into a new statement/predicate using either the "and" or the "or" connectors. For example, consider the two statements "it is raining" and "I have my umbrella up", we can combine these as follows:

"It is raining AND I have my umbrella up".

"It is raining OR I have my umbrella up".

Instead of using the words "and" and "or", it is often easier to use the mathematical notation $\wedge$ to mean 'and' and $\vee$ to mean 'or'. Hence, $A \wedge B$ is the same as A and B, whereas, $A \vee B$ is the same as A or B.

Table 3.2(a) and (b) are the truth tables for the $\wedge$ and $\vee$ connectors, respectively. For example, if both A and B are true, then $A \wedge B$ is also true. We remark that if both A and B are true, then $A \vee B$ is also true, it does not have to be only one of them: "or" is the *inclusive* "or".

A	B	$A \wedge B$		A	B	$A \vee B$
T	T	T		T	T	T
T	F	F		T	F	T
F	T	F		F	T	T
F	F	F		F	F	F
	(a)				(b)	

Table 3.2: Truth tables for (a) the and connector, (b) the or connector.

Example 3.1

Let $p(x)$ be the predicate $x \geq 5$ and $q(x)$ be the predicate $x < 12$. Then the predicate "$p(x)$ and $q(x)$" is equivalent to

$$5 \leq x < 12.$$

Example 3.2

Let $p(x)$ be the predicate $x < 0$ and $q(x)$ be the predicate $x = 0$. Then the predicate "$p(x)$ or $q(x)$" is equivalent to

$$x \leq 0.$$

When we negate compound statements involving $\wedge$ and $\vee$, the roles of $\wedge$ and $\vee$ are switched, so that

- $\neg(A \wedge B)$ is equivalent to $\neg A \vee \neg B$,
- $\neg(A \vee B)$ is equivalent to $\neg A \wedge \neg B$.

Some written examples are:

- $\neg$ "it is raining AND I have my umbrella up" means "It is NOT raining OR I do NOT have my umbrella up".
- $\neg$"It is raining OR I have my umbrella up" means "It is NOT raining AND I do NOT have my umbrella up".

Below are some more mathematical examples.

- Consider the predicate "$x < 2$ or $x > 4$". Negating the predicate, we obtain "$x \geq 2$ and $x \leq 4$", *i.e.* $2 \leq x \leq 4$.
- The negation of the predicate "x is even and x is a multiple of 4" is "x is odd or x is not a multiple of 4".

.3 "There Exists" and "For All"

Statements and predicates, especially mathematical ones, often include phrases involving "all" and "there exists". For example, we might have the statements:

- A: All cats have four paws.
- B: There exists a prime number which is even.

A is a false statement, but B is true. As statements such as these are so ubiquitous, we have the special mathematical notation $\forall$, which means "for all", "for any" or just "all". Similarly, we have $\exists$ which means "there exists", or "there is".

What is the negation of statement A? One way is to simply say "Not all cats have four paws", but equivalently we could have said "There exists a cat which does not have four paws' and we notice that the "all" has been replaced with "there exists" and the negation of "have four paws", "does not have four paws", has been used.

Similarly, let us negate statement B. "Not there exists a prime number which is even" does not really make sense, so alternatively we could have said "There does not exist a prime number which is even". This is equivalent to saying "All prime numbers are odd". In this case, the "there exists" has been changed to "all" and the negative has been carried through so "which is even" has become "are odd". We summarise this below.

- $\neg(\forall A)$ is equivalent to $\exists(\neg A)$.
- $\neg(\exists A)$ is equivalent to $\forall(\neg A)$.

With these simple rules, we can take quite complex expressions and understand how to negate them.

Example 3.3

Let $p(x)$ and $q(x, y)$ be predicates, negate the statement:

$$\forall x \; \exists y \; p(x) \vee q(x, y).$$

If $p(x)$ means $x > 2$ and $q(x, y)$ means $x > y$, rewrite both the statement above and its negation. Are these statements true or false?

Solution:

Using the laws stated above, we find the negation is

$$\exists x \; \forall y \; \neg p(x) \wedge \neg q(x, y).$$

For the predicates given, the original statement reads: "For all x, there is a y such that $x > 2$, or $x > y$". Note, this is a true statement, if we pick $y = x - 1$, then $x > y$ is satisfied.

Similarly, the negation reads: "There exists an x such that, for all y, $x \leq 2$ and $y \leq x$". This statement is false, as, for any $x < 2$, we can find a $y > x$.

Example 3.4

Negate the statement: "In every country there is a city where every resident either rides a bike or drives a car but does not take the bus".

Solution:

To make our lives easier, let A denote "country", B denote "city", C denote "resident", D denote "rides a bike", E denote "drives a car" and F denote "takes the bus". Then the original statement can be written as statement P:

$$P \colon \forall A \; \exists B \; \forall C \; (D \vee E) \wedge \neg F.$$

Applying the laws of negation, we find

$$\neg P \colon \exists A \; \forall B \; \exists C \; (\neg D \wedge \neg E) \vee F.$$

We can decode this to find the negation of the original statement is: There exists a country where in all cities there is a resident who either takes the bus or who neither rides a bike nor drives a car.

Exercise 3.1

Q1. Which of the following are statements, which are predicates and which are neither? If they are statements, are they true or false?

    (a) All dogs have a tail;

    (b) There is a donkey;

    (c) A monkey;

    (d) θ;

    (e) $\cos(\theta)$;

    (f) $\cos(\theta) = 0$;

    (g) $1 = 1$;

    (h) $1 = 2$;

    (i) $x = 1$;

    (j) q implies p;

    (k) If q implies p and q is true, then p is true;

    (l) If q implies p, then p implies q;

Q2. Write in words the negation of the following statements:

    (a) All cars have a handbrake;

    (b) All cars have a handbrake that is guaranteed to stop the car;

    (c) There is a flight where smoking is allowed in every seat;

    (d) In every town there is a restaurant where all items on the menu contain potatoes.

    (e) Everybody who buys *Stoves Today* either reads it, or uses it for kindling.

Q3. Negate the following mathematical statements:

    (a) $\forall x \; \neg p(x)$;

    (b) $\exists x \; \neg p(x)$;

    (c) $\exists x \; \forall y \; p(x, y)$;

    (d) $\forall x \; \exists y \; \exists z \; p(x, y, z)$;

    (e) $\exists y \; \forall x \; \forall z \; \neg p(x, y, z)$;

    (f) $\exists x \; \forall y \; (p(x) \vee q(y))$;

    (g) $\forall z \; \forall y \; \forall x \; (p(x) \wedge (q(y) \vee \neg r(z)))$;

    (h) $(\exists x \; p(x)) \wedge (\neg \forall y \; p(y))$.

3.2 Implication

Statements are connected using *implication arrows* '$\Rightarrow$', '$\Leftarrow$' and '$\Leftrightarrow$' to form *compound statements*.

3.2.1 One-way Implication

Definition 3.1 — Implication

If we write $A \Rightarrow B$ then this is the compound statement "A implies B", where A and B are statements. This is equivalent to the compound statement "if A then B". If A holds, then B must also hold.

Example 3.5

Applying Definition 3.1, we see that $x < 0 \Rightarrow x \leq 0$ is a true statement, since any number strictly less than zero must also be less than or equal to zero.

Example 3.6

The statement

$$x = 1 \quad \Rightarrow \quad x^2 = x,$$

for x a real number, is true. The second statement follows directly from the first by multiplication of both sides by x.

Remark

The statement $A \Rightarrow B$ can also be written as "A only if B", *i.e.* A can only hold if B holds. As an example, consider the true statement $x \geq 3 \Rightarrow x \geq 2$, in words this says "if x is greater than 3 then x is greater than 2". However, we could also say "x is greater than 3 only if x is greater than 2".

Remark

The statement $B \Rightarrow A$ is equivalent to $A \Leftarrow B$.

We can produce more detailed statements by combining simple statements using "and" and "or".

- The statement "if A and B then C" means that C holds if **both** A and B hold.
- The statement "if A or B then C" means that C holds if **at least one** of A or B holds.

3.2.2   Converse and Contrapositive

It is very important to note that "if A then B" is *not* the same as "A if B". When we write "A if B", this means "if B then A", *i.e.* $B \Rightarrow A$. $B \Rightarrow A$ is known as the *converse* of $A \Rightarrow B$.

Definition 3.2 — Converse

If $A \Rightarrow B$, then the *converse* of this statement is

$$B \Rightarrow A.$$

If the statement $A \Rightarrow B$ is true, the converse ($B \Rightarrow A$) is not necessarily true, as we see in the following example.

Example 3.7

The statement

$$x = 1 \quad \Rightarrow \quad x^2 = x,$$

is true. Its converse is

$$x^2 = x \quad \Rightarrow \quad x = 1.$$

This statement is false, since $x = 0$ also satisfies $x^2 = x$.

Definition 3.3 — Contrapositive

If $A \Rightarrow B$, then the *contrapositive* of this statement is

$$\neg B \Rightarrow \neg A.$$

Example 3.8

Consider the statement $x > 3 \Rightarrow x^2 > 9$, then the contrapositive is

$$x^2 \leq 9 \Rightarrow x \leq 3.$$

We notice that both the original and the contrapositive are both true.

The result from Example 3.8 is true in general: $A \Rightarrow B$ and its contrapositive are *always* equivalent statements. We shall see in Chapter 5 that, if we wish to prove $A \Rightarrow B$, then we can instead prove the contrapositive and use the equivalence result to show $A \Rightarrow B$. This is an example of an *indirect proof*.

Exercise 3.2

Show, with a truth table, that $A \Rightarrow B$ is logically equivalent to $\neg B \Rightarrow \neg A$.

Example 3.9

Which of the following statements is the odd one out?

    A: All carrots are vegetables;
    B: There is no carrot which is not a vegetable;
    C: If there is anything which is not a carrot, then it is not a vegetable.
    D: If there is anything which is not a vegetable, then it is not a carrot.

Solution:

To answer this question, we can rewrite each expression in a more mathematical way. Let C denote carrots, V denote vegetables, then the statements can be rewritten as

    A: $C \Rightarrow V$;
    B: $\neg \exists C, \neg V$;
    C: $\neg C \Rightarrow \neg V$;
    D: $\neg V \Rightarrow \neg C$.

Statement D is clearly the contrapositive of A, hence they are equivalent to each other. Statement C is the converse of A and so they are not equivalent. The more difficult statement is B, but recalling the rules of negation, this can be rewritten as $\forall C$, V, which in turn is equivalent to $C \Rightarrow V$. Thus, statement C is the odd one out. We also remark that C is a false statement (clearly there are vegetables which are not carrots), whereas, the other statements are true.

3.2.3 Two-way Implication

If the statements $A \Rightarrow B$ and $B \Rightarrow A$ are both true, then we have a two-way implication.

Definition 3.4 — Two-way implication
If $A \Rightarrow B$ and $A \Leftarrow B$, we write

$$A \Leftrightarrow B.$$

This is often expressed as "A if and only if B". Recall that $A \Leftarrow B$ is "A if B" and

$A \Rightarrow B$ is "A only if B".

Remark
For the statement $A \Leftrightarrow B$ we often write A iff B, where iff is short for *if and only if*.

Example 3.10
The statement

$$x + 5 = 3 \quad \Leftrightarrow \quad x = -2,$$

is a two-way implication, since the statements

$$x + 5 = 3 \quad \Rightarrow \quad x = -2,$$

and

$$x = -2 \quad \Rightarrow \quad x + 5 = 3,$$

are both true.

3.2.4 Strength of Statements

For statements A and B, if A implies B, but B does *not* imply A, then statement A is *stronger* than statement B. The strongest possible compound statement we can write is then $A \Rightarrow B$. However, if A implies B *and* B implies A, then the strongest possible compound statement is $A \Leftrightarrow B$.

Example 3.11

Let the statement A be $x > 2$ and the statement B be $x > 0$ for $x \in \mathbb{R}$.

We have $A \Rightarrow B$, since all real numbers greater than 2 are also greater than 0. However, $B \nRightarrow A$, since not all real numbers greater than 0 are also greater than 2, *e.g.* $x = 1$. Hence, A is a stronger statement than B and the strongest compound statement we can write is $A \Rightarrow B$.

Exercise 3.3

Q1. Insert one of the symbols $\Rightarrow$, $\Leftarrow$, $\Leftrightarrow$ to make the strongest possible compound statement from each of the following, where n is an integer and x is a real number in each case.

   (a) S is a rhombus _____ S is a square.
   (b) X is a rhombus _____ X is a parallelogram.
   (c) $n^2 > 8$ and $n > 0$ _____ $n \geq 3$.
   (d) x^2 is rational _____ x is rational.

Q2. This question is about constructing sensible mathematical arguments. In each case, does the conclusion Q follow logically from the premises P1 and P2? Here, x, y, z, where they occur, are real numbers.

   (a)   P1:   All cats have wings.
          P2:   All winged creatures have four legs.
          Q: All cats have four legs.
   (b)   P1:   If $x^2 + y^2 \leq 1$ then $-1 \leq x \leq 1$.
          P2: $4x = 1$.
          Q: $x^2 + y^2 \leq 1$.
   (c)   P1: $y \leq 4$ only if $x \geq 3$.
          P2:   If $y \leq 4$ then $z^2 > x$.
          Q: If $z = 0$ then $y > 0$.
   (d)   P1:   If $x > 0$ then $y > 5$.
          P2:   If $x \leq 0$ then $z < 7$.
          Q: $y > 5$ only if $z \geq 7$.
   (e)   P1: $y > 3$ only if $x^2 > 4$.
          P2:   If $y > 3$ then $z \geq 5$.
          Q: If $y > 3$ then $z > x^2$.

Q3. Citizens living in the faraway republic of Mathtopia organise their hobbies according to a very strict set of rules. In Mathtopia, it is forbidden to write or speak in ambiguous terms, so the rules are written using mathematical language to guarantee no ambiguity! The rules are as follows:

   (a) No Mathtopian that eats bagels cannot swim.
   (b) No Mathtopian without a calculator plays poker.
   (c) Mathtopians who have an abacus all eat bagels.
   (d) No Mathtopian who can swim likes fencing.
   (e) No Mathtopian has a calculator unless they have an abacus.

Using these 5 rules, what conclusion can we legitimately draw about citizens of Mathtopia?

Hint: Label the various properties (e.g. B for "eats bagels") and try to find a chain of implications.
Use $\neg A$ to indicate "not A".
Note that "A unless B" means $\neg B \Rightarrow A$.

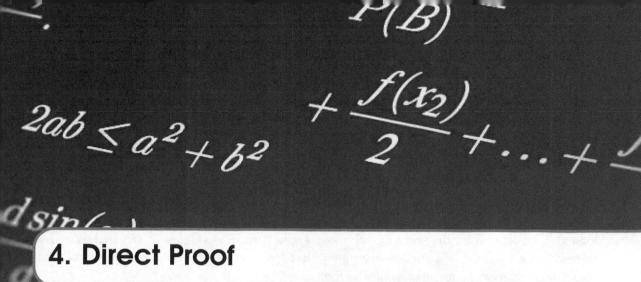

4. Direct Proof

As mentioned in the introduction, proof is one of the cornerstones of mathematics. This is what sets mathematics apart from the other sciences. Once a mathematical result has been proved it holds for evermore; it is not subject to revision in the light of new experimental evidence. In 1637, the French mathematician Pierre de Fermat conjectured that there were no positive integer solutions of the equation $a^n + b^n = c^n$, for $n > 2$.

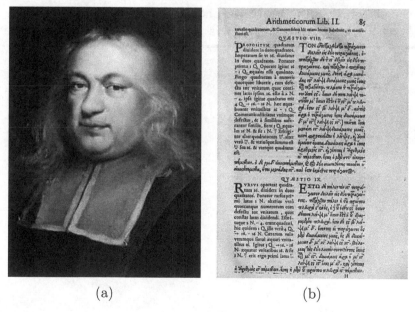

(a)   (b)

Figure 4.1: Pierre de Fermat (a) and the original problem against which he wrote his conjecture.

He made this statement in the margin of his copy of Diophantus' Arithmetica, next to Problem II.8 and claimed "It is impossible to separate a cube into two cubes, or a fourth power into two fourth powers, or in general, any power higher than the second, into two like powers. I have discovered a truly marvellous proof of this, which this margin is too narrow to contain."

The conjecture that no such solutions exist for $n > 2$ remained unproven until 1994, when the English mathematician Andrew Wiles released a proof.

> **Remark**
>
> For $n = 2$ we know that there exist infinitely many positive integer solutions to the equation $a^2 + b^2 = c^2$; these are Pythagorean triples.

A *direct proof* is a process where we try and reach a conclusion by following a series of logical steps, starting from some initial result(s). In turn, the conclusion reached may then be used as a premise in a further argument that leads to yet another conclusion. This is sometimes referred to as *proof by deduction*.

4.1   Examples of Direct Proof

We consider some motivating examples that should be familiar from studies at GCSE level.

> **Example 4.1**
> Prove that $(2x - 1)(x + 3) = 2x^2 + 5x - 3$ for all real x.
>
> **Solution:**
> Expanding the brackets on the left-hand side, we obtain
>
> $$(2x - 1)(x + 3) = 2x^2 + 6x - x - 3 = 2x^2 + 5x - 3.$$
>
> Since we have shown that the left and right-hand sides of the identity are equivalent, we have proven directly that the identity is true for all real x.

Example 4.2

Given that the interior angles of a convex quadrilateral sum to 360°, use the diagram below to prove that corresponding angles are of equal size.

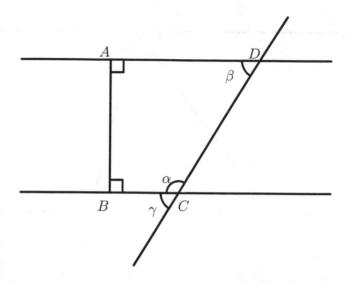

Solution:

Clearly, $ABCD$ is a quadrilateral, so its interior angles sum to 360°. Since angles DAB and ABC are right-angles, $\alpha + \beta = 360° - 180° = 180°$. Furthermore, angles α and γ lie along a straight line, so $\alpha + \gamma = 180°$. Thus, $\alpha + \beta = \alpha + \gamma = 180° \Rightarrow \beta = \gamma$, so the corresponding angles a and c are equal.

Example 4.3

Prove that the sum of any four consecutive integers is even.

Solution:

We let the first number be n, and so the second, third and fourth numbers are $n + 1$, $n + 2$, $n + 3$ respectively. Adding these,

$$n + (n + 1) + (n + 2) + (n + 3) = 4n + 6,$$
$$= 2(2n + 3).$$

Since this is a multiple of two, the sum of any four consecutive integers is even.

Example 4.4

Consider the diagram below.

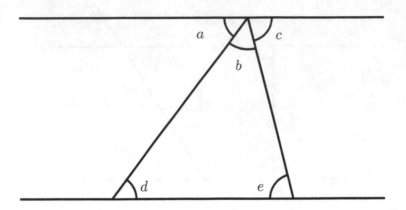

Use properties of parallel lines to prove that the interior angles of a triangle sum to 180°.

Solution:

Firstly, angles a and d are alternate angles, so $a = d$. Similarly, c and e are alternate, so $c = e$. Secondly, a, b and c together form a straight line, so $a+b+c = 180°$. Combining these facts, we obtain $b + d + e = 180°$. Since none of the angles were specified, this equation holds for all possible angles and therefore all possible triangles. Hence, the interior angles of all triangles sum to 180°.

Example 4.5

Consider

$$y = \frac{7x - 1}{4x + 3}, \quad x \neq -\frac{3}{4}.$$

(a) Prove directly that if $x > \frac{1}{2}$, then $y > \frac{1}{2}$ also.

(b) Show further that if $x > \frac{1}{2}$, $x > y$.

Solution:

(a) Suppose that $x > \frac{1}{2}$. Consider $y - \frac{1}{2}$:

$$y - \frac{1}{2} = \frac{7x - 1}{4x + 3} - \frac{1}{2} = \frac{14x - 2 - (4x + 3)}{2(4x + 3)} = \frac{10x - 5}{2(4x + 3)} = \frac{5(2x - 1)}{2(4x + 3)}.$$

Since $x > \frac{1}{2}$, $2x - 1 > 0$ and $4x + 3 > 0$, so $y - \frac{1}{2} > 0$.

Hence, $x > \dfrac{1}{2} \Rightarrow y - \dfrac{1}{2} > 0 \Rightarrow y > \dfrac{1}{2}$, as required.

(b) Suppose again that $x > \dfrac{1}{2}$. Now consider $x - y$:

$$x - y = x - \frac{7x - 1}{4x + 3} = \frac{4x^2 + 3x - (7x - 1)}{4x + 3} = \frac{4x^2 - 4x + 1}{4x + 3} = \frac{(2x - 1)^2}{4x + 3}.$$

Since $x > \dfrac{1}{2}$, $(2x - 1)^2 > 0$ and $4x + 3 > 0$, so $x - y > 0$.

Hence, $x > \dfrac{1}{2} \Rightarrow x - y > 0 \Rightarrow x > y$, as required.

Proof Tip

Initial assumptions and known facts should be stated clearly before starting the proof. Once this has been done, ensure that a written proof is concise and easy to follow.

Learning Resource 4.1 — Structuring Proofs

From the digital book an activity can be downloaded where a framework for structuring proofs is explained.

Exercise 4.1

Q1. Prove that $(x + y)(x - y) = x^2 - y^2$ for all real x and y.

Q2. Prove that if $(x - p)$ is a factor of the polynomial expression $f(x)$ then $f(p) = 0$.

Q3. Prove that $x^3 + x^2 - 2x - 8$ only has one real factor.

Q4. The equation $kx^2 + 2kx - 3 = 0$, where k is a constant, has no real roots. Prove that k satisfies the inequality $-3 < k \leq 0$.

Q5. For which values of the constant k does the equation $kx^2 - 3x + k = 0$ have two distinct real roots?

Q6. (a) Prove that the interior angles of a quadrilateral sum to $360°$.

(b) Prove that the interior angles of an n-sided polygon sum to $180(n - 2)°$.

Q7. Prove that $x^2 - 6x + 10 \geq 1$ for all real x.

Q8. The line with equation $y = mx + 2$ intersects the circle with equation $(x + 1)^2 + (y - 2)^2 = r$, $r > 0$, at two distinct points. Prove that $r > \dfrac{m^2}{1 + m^2}$.

Q9. Prove that the function $f(x) = 3x - 1$ crosses the x-axis exactly once.

Q10. Prove that the function $f(x) = x^2 - 5$ crosses the x-axis exactly twice.

Q11. Prove that the function $f(x) = x^3 - x - 1$ has a root between $x = 1$ and $x = 2$. Show further that $f(x)$ has no other real roots.

Q12. Consider the function

$$f(x) = \frac{\sqrt{a+x} - \sqrt{a}}{x}, \quad a > 0.$$

Prove directly that $\lim\limits_{x \to 0} f(x) = \frac{1}{2\sqrt{a}}$.

Q13. Consider the expression a_n, where

$$a_n = \frac{2n - \sqrt{n}}{n}, \quad n \in \mathbb{N}.$$

Prove directly that $a_n < 2$ for all n.

Q14. By finding an example, prove that there exist irrational a and b such that a^b is rational.

Hint: The fact that $\sqrt{2}$ is irrational may be used without proof.

Q15. Let $a, b, c, d \in \mathbb{R}$, with $a < b$, $c < d$. Prove or disprove each of the following inequalities.

(a) $a + c < b + d$.

(b) $a - c < b - c$.

(c) $ac < bd$.

(d) $\dfrac{a}{c} < \dfrac{b}{d}$ $(c \neq 0, d \neq 0)$.

(e) $a - \dfrac{1}{c} < b - \dfrac{1}{d}$ $(c \neq 0, d \neq 0)$.

(f) $a^3 < b^3$.

(g) $a^3 - d^3 < b^3 - c^3$.

Q16. Find all real values of k for which the inequality

$$\frac{x - 1}{x^2 + 2x + 3} \leq \frac{x + k}{x^2 + 4x + 5}$$

holds for all $x \in \mathbb{R}$.

Q17. Consider the diagram below.

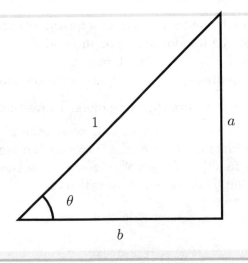

Use this diagram to provide a geometric proof of the identity

$$\cos(\arcsin(x)) = \sin(\arccos(x)) = \sqrt{1 - x^2}.$$

Q18. (a) Use the fact that, for $n \in \mathbb{N}$, real numbers $a_1, \ldots, a_n, b_1, \ldots, b_n$, and $x \in \mathbb{R}$,

$$F(x) = \sum_{k=1}^{n}(xa_k - b_k)^2 \geq 0,$$

to prove the Cauchy-Schwarz inequality:

$$\left(\sum_{k=1}^{n} a_k b_k\right)^2 \leq \left(\sum_{k=1}^{n} a_k^2\right)\left(\sum_{k=1}^{n} b_k^2\right).$$

(b) Use the Cauchy-Schwarz inequality to prove the arithmetic-harmonic mean inequality:

$$\frac{x_1 + \ldots x_n}{n} \geq \frac{n}{\frac{1}{x_1} + \ldots + \frac{1}{x_n}}.$$

This inequality states that the mean of the positive numbers $x_1, \ldots, x_n$ is greater than or equal to the reciprocal of the mean of their reciprocals.

Hint: Apply Cauchy-Schwarz to $\left(\sum_{k=1}^{n} 1\right)^2$.

Q19. For positive real numbers x and y, it is a fact that

$$(\sqrt{x} - \sqrt{y})^2 \geq 0.$$

(a) Use this fact to deduce the arithmetic-geometric mean inequality for two positive real numbers.

(b) Suppose now that a, b and c are positive real numbers which satisfy $a+b+c = 1$. Prove that the *minimum* value of

$$\left(\frac{1-a}{a}\right)\left(\frac{1-b}{b}\right)\left(\frac{1-c}{c}\right)$$

is 8.

.2 Proof by Exhaustion

We often consider statements for which there are a finite number of cases. We can then perform a direct calculation for each case to deduce that the statement holds in all possible cases.

Example 4.6

Prove that for every positive integer n, where $3 \leq n \leq 8$ that the positive integer $n^2 + 3n$ is even.

Solution:
We consider each possible n.

$$n = 3 : \quad 3^2 + 3 \times 3 = 18,$$
$$n = 4 : \quad 4^2 + 3 \times 4 = 28,$$
$$n = 5 : \quad 5^2 + 3 \times 5 = 40,$$
$$n = 6 : \quad 6^2 + 3 \times 6 = 54.$$
$$n = 7 : \quad 7^2 + 3 \times 7 = 70.$$
$$n = 8 : \quad 8^2 + 3 \times 8 = 86.$$

These are all even, and so the statement is true as it has been verified for every value of n directly.

Remark

The above statement is in fact true for all integer n.

Example 4.7

Prove that for every positive integer n, where $7 \leq n \leq 14$, the positive integer $n^2 - n - 1$ is either prime or the product of two primes.

Solution:
Considering each possible n,

$$n = 7 : \qquad 7^2 - 7 - 1 = 41 \quad \text{(prime)},$$
$$n = 8 : \quad 8^2 - 8 - 1 = 55 = 5 \times 11 \quad \text{(product of two primes)},$$
$$n = 9 : \qquad 9^2 - 9 - 1 = 71 \quad \text{(prime)},$$
$$n = 10 : \quad 10^2 - 10 - 1 = 89 \quad \text{(prime)},$$
$$n = 11 : \quad 11^2 - 11 - 1 = 109 \quad \text{(prime)},$$
$$n = 12 : \quad 12^2 - 12 - 1 = 131 \quad \text{(prime)},$$
$$n = 13 : \quad 13^2 - 13 - 1 = 155 = 5 \times 31 \quad \text{(product of two primes)},$$
$$n = 14 : \quad 14^2 - 14 - 1 = 181 \quad \text{(prime)}.$$

We have verified the statement for all possible values of n directly. Hence, the statement is true.

Remark

A famous example where proof by exhaustion was used is the *Four Colour Theorem*. The question asks whether it is possible to use only four colours to colour a map so that no adjacent countries have the same colour.

Figure 4.2: A map of Europe using only four colours to shade the countries.

Initially, more complex mathematics was used to reduce the problem to a finite, but large number of cases, from which a proof by exhaustion argument could follow. There were far too many cases to be considered by hand, so a computer was employed to check all the cases. This proof was one of the first to be carried out using a computer. The Four Colour Theorem has further uses in more abstract areas of mathematics.

Example 4.8

Prove that all cube numbers are either a multiple of 9, or 1 more, or 1 less than a multiple of 9.

Solution:

All positive integers are either a multiple of 3, or one more or one less than a multiple of 3. We consider each case individually.

- n is a multiple of 3, *i.e.* there exists positive integer k such that $n = 3k$. Then $n^3 = 27k^3 = 9 \times 3k^3$, so n^3 is a multiple of 9.
- n is one more than a multiple of 3, *i.e.* there exists positive integer k such that $n = 3k + 1$. Then $n^3 = 27k^3 + 27k^2 + 9k + 1 = 9k(3k^2 + 3k + 1) + 1$, so n^3 is one more than a multiple of 9.
- n is one less than a multiple of 3, *i.e.* there exists positive integer k such that $n = 3k - 1$. Then $n^3 = 27k^3 - 27k^2 + 9k - 1 = 9k(3k^2 - 3k + 1) - 1$, so n^3 is one less than a multiple of 9.

Since we have found the statement consistent in all possible cases, we have proven that all cube numbers are either a multiple of 9 or 1 more or 1 less than a multiple of 9.

Example 4.9

Prove that $x = n^3 - n$ is divisible by 12 for all odd integers $n > 2$.

Solution:

Factorising gives

$$n^3 - n = n(n^2 - 1) = (n-1)n(n+1),$$

so x is the product of 3 consecutive positive integers.

Since n is odd, $(n-1)$ and $(n+1)$ are even, which means that x is divisible by 4 for all n. We then have two cases:

- n is a multiple of 3, so x is divisible by $3 \times 4 = 12$.
- n is not a multiple of 3. However, exactly one of any 3 consecutive integers is a multiple of 3, so one of $(n-1)$ and $(n+1)$ is a multiple of 3, so x is divisible by $3 \times 4 = 12$.

Since we have found that the statement is true for all possible cases, it is true for all odd $n > 2$.

4.3   Proofs of Statements Involving the Positive Integers

Many proofs involving positive integers require us to consider odd and even numbers separately, or the statement itself may only concern odd (or even) numbers. Because of this, we can use some basic algebra to write:

- The positive integer m is *even* if there exists a positive integer p such that $m = 2p$,
- The positive integer n is *odd* if there exists a positive integer q such that $n = 2q - 1$.

Example 4.10

Prove directly that

    (a) the sum of two even positive integers is even,

    (b) the sum of two odd positive integers is even.

Solution:

    (a) Let m and n be even numbers, *i.e.* there exist positive integers p and q such that $m = 2p$ and $n = 2q$. Then

$$m + n = 2p + 2q = 2(p + q).$$

Since $p + q$ is a positive integer, $m + n$ has the form of an even number. Hence, the sum of two even numbers is even.

    (b) Let m and n be odd numbers, *i.e.* there exist positive integers p and q such that $m = 2p - 1$ and $n = 2q - 1$. Then

$$m + n = 2p - 1 + 2q - 1 = 2p + 2q - 2 = 2(p + q - 1).$$

Since $p + q - 1$ is an integer, $m + n$ has the form of an even number. Hence, the sum of two odd numbers is even.

Example 4.11

Prove that the product of two consecutive odd numbers is 1 less than a multiple of 4.

Solution

We can express two consecutive odd numbers m and n as $m = 2k - 1$ and $n = 2k + 1$, where k is a positive integer. Multiplying,

$$mn = (2k - 1)(2k + 1) = 4k^2 - 1.$$

Hence, the product of two consecutive odd numbers is one less than a multiple of 4.

.4 Disproof by Counter Example

To prove that a statement is false, it is sufficient to find just one counter example.

Example 4.12

A student is trying to prove the following inequality for all integers x and y,

$$(x + y)^4 \leq x^4 + y^4.$$

Explain why they will not be able to prove this to be true.

Solution:

If we consider $x = 0, y = 1$, then,

$$(x + y)^4 = (0 + 1)^4,$$
$$= 1,$$
$$x^4 + y^4 = 0^4 + 1^4,$$
$$= 1.$$

However, considering $x = 1, y = 2$, we have

$$(x + y)^4 = (1 + 2)^4,$$
$$= 81,$$
$$x^4 + y^4 = 1^4 + 2^4,$$
$$= 17.$$

Hence, we have found a counter example, disproving the statement. A student cannot possibly prove a statement to be true if there exists a counter example.

Example 4.13

A Mersenne prime is a prime number that can be written in the form $2^n - 1$ for some positive integer $n > 1$. Prove that the statement

"For every prime number p, the number $2^p - 1$ is a Mersenne prime"

is false.

Solution:

If we consider the first few prime numbers, we obtain

$$2^2 - 1 = 3, \quad 2^3 - 1 = 7, \quad 2^5 - 1 = 31.$$

It appears that the statement is true so far! However, we have only tried three cases, where there are in fact infinitely many choices of p. We only need to show that the statement fails for **one** choice of p to disprove the statement. Considering $p = 11$, we obtain

$$2^{11} - 1 = 2047 = (23)(89).$$

Hence, we have found a counterexample, disproving the statement.

Exercise 4.2

Q1. Find a counter example to disprove the following statements.
   (a) For all positive integers $n \leq 10$, $n! < 3^n$.
   (b) All non-prime integers have an even number of distinct prime factors.
   (c) $n^2 + 2n - 2$ is even for all positive integer values of n.
   (d) All straight lines of the form $y = mx + c$ intersect the x-axis.

Q2. (a) Prove that for all positive integer values of a and b,

$$\frac{a}{b} + \frac{b}{a} \geq 2.$$

   (b) Find a counter example to disprove this statement when at least one of a or b is negative.

Q3. Prove the following statements about positive integers directly.
   (a) If n is odd, then n^2 is odd.
   (b) If m and n are odd, then mn is also odd.
   (c) If m and n are even, then mn is also even.
   (d) If m is odd and n is even, then mn is even.
   (e) If n is even, then $7(n + 4)$ is even.

Q4. For each of the statements in Q3, write down the converse and either prove that the new statement is true or provide a counter example to show that it is false.

Q5. Prove that the sum of the squares of two positive integers is less than the square of the sum of the numbers.

Q6. For the following statements, either prove the statement directly or find a counter example to disprove it.

(a) The difference of the squares of two consecutive even numbers is a multiple of 4.

(b) The difference of the squares of two consecutive odd numbers is a multiple of 4.

(c) The difference of the squares of two consecutive numbers is a multiple of 4.

Q7. Find a counter example to disprove the statement that for positive integer $n \geq 3$, $n! - 1$ is prime.

Q8. (a) Let a, b and c be positive integers. Prove that if b and c are divisible by a, then $b + c$ is also divisible by a.

(b) Let a, b and c be positive integers. Prove that if b is divisible by a and c is divisible by b, then c is divisible by a.

Q9. Do there exist integers a, b and c such that

$$4a + 3 = b^2 + c^2?$$

Q10. (a) Find all possible rectangles, having sides of positive integer length, which have the property that the perimeter is equal (numerically) to the area. It is not enough to identify some examples with this property; prove that there are no more solutions that have been overlooked.

Hint: The side-lengths can be labelled as x and y in such a way that $x < y$. Obtain a relationship between x and y and think about $1/x$ and $1/y$. How large can x be?

(b) Find all possible cuboids, having sides of positive integer length p, q and r, for which the total surface area (numerically) is twice the volume.

Hint: Similar to the first problem, assume $p \leq q \leq r$. Consider $1/p$, $1/q$, $1/r$ and how large p can be.

5 Direct Proofs From A-Level Mathematics

In this section we collect together some direct proofs of results used in A-Level Mathematics as examples.

Example 4.14

Write out a simple proof for each of the index laws below

$$b^m \times b^n = b^{m+n}, \tag{4.1}$$

$$b^m \div b^n = b^{m-n}, \tag{4.2}$$

$$(b^m)^n = b^{mn}, \tag{4.3}$$

$$b^0 = 1, \qquad b \neq 0, \tag{4.4}$$

$$b^{-m} = \frac{1}{b^m}, \qquad b \neq 0. \tag{4.5}$$

Solution:

We can prove Property (4.1) simply by using the definition of the notation as follows:

$$b^m \times b^n = \underbrace{\overbrace{b \times b \times \cdots \times b}^{m \text{ times}} \times \overbrace{b \times b \times \cdots \times b}^{n \text{ times}}}_{m + n \text{ times}} = b^{m+n}.$$

For property (4.2) we have n lots of b which cancel out leaving $m - n$ lots of b.

$$b^m \div b^n = \frac{\overbrace{b \times b \times b \times b \times \cdots \times b}^{m \text{ times}}}{\underbrace{b \times b \times \cdots \times b}_{n \text{ times}}} = b^{m-n}$$

Property (4.3) is simply n brackets each containing m lots of b.

$$(b^m)^n = \underbrace{(\overbrace{b \times b \times \cdots \times b}^{m \text{ times}}) \times (\overbrace{b \times b \times \cdots \times b}^{m \text{ times}}) \times \cdots \times (\overbrace{b \times b \times \cdots \times b}^{m \text{ times}})}_{n \text{ brackets}} = b^{mn}$$

We can think of property (4.4) as $b^m \div b^n$ where $m = n$. As a fraction, each b will cancel out leaving 1 as the simplest form. This also illustrates why b should not be zero as division by zero is undefined.

$$b^0 = b^{m-m} = \frac{b^m}{b^m} = 1$$

Finally, to prove property (4.5) we can apply property (4.2) and property 4.4.

$$b^{-m} = b^{0-m} = \frac{b^0}{b^m} = \frac{1}{b^m}$$

Example 4.15

Prove the following properties.

$$\sqrt{x} \times \sqrt{y} = \sqrt{x \times y}; \tag{4.6}$$

$$\sqrt{x} \div \sqrt{y} = \sqrt{\frac{x}{y}}; \tag{4.7}$$

$$\sqrt{x + y} \leq \sqrt{x} + \sqrt{y} \qquad \forall x, y \geq 0. \tag{4.8}$$

Solution:

Property (4.6) can be proved in the following way. Let $a = \sqrt{x} \times \sqrt{y}$ and $b = \sqrt{x \times y}$. Squaring these we obtain,

$$a^2 = (\sqrt{x}\sqrt{y})^2 = (\sqrt{x}\sqrt{y})(\sqrt{x}\sqrt{y}) = \sqrt{x}\sqrt{x}\sqrt{y}\sqrt{y} = xy$$
$$b^2 = (\sqrt{xy})^2 = xy.$$

Equating these we have $a^2 = b^2$. Since $x, y \geq 0$ we also have $a, b \geq 0$. Hence $a = b$ and

$\sqrt{x}\sqrt{y} = \sqrt{xy}.$

For (4.7), let $a = \frac{\sqrt{x}}{\sqrt{y}}$ and $b = \sqrt{\frac{x}{y}}$.

$$a^2 = \left(\frac{\sqrt{x}}{\sqrt{y}}\right)^2 = \frac{\sqrt{x}}{\sqrt{y}} \times \frac{\sqrt{x}}{\sqrt{y}} = \frac{x}{y}$$

$$b^2 = \left(\sqrt{\frac{x}{y}}\right)^2 = \frac{x}{y}.$$

Since $x, y \geq 0$ we have $a, b \geq 0$, hence,

$$a^2 = b^2$$
$$\Rightarrow \quad a = b$$
$$\Rightarrow \quad \frac{\sqrt{x}}{\sqrt{y}} = \sqrt{\frac{x}{y}}.$$

To prove Property (4.8) we begin by noting that for $a, b \geq 0$ the inequality $a \leq b \iff a^2 \leq b^2$ holds. Apply this to the triangle inequality.

$$\sqrt{x+y} \leq \sqrt{x} + \sqrt{y}$$
$$\Leftrightarrow \quad \left(\sqrt{x+y}\right)^2 \leq \left(\sqrt{x} + \sqrt{y}\right)^2$$
$$\Leftrightarrow \quad x + y \leq x + y + 2\sqrt{xy}.$$

Since $x, y \geq 0$ we have $2\sqrt{xy} \geq 0$, so the inequality must hold.

Example 4.16
Laws of Logarithms
Let $x > 0$, $y > 0$ and base $a > 0$, prove the following results for logarithms:

- **Multiplication Law**

$$\log_a(x) + \log_a(y) = \log_a(xy). \tag{4.9}$$

- **Division Law**

$$\log_a(x) - \log_a(y) = \log_a\left(\frac{x}{y}\right). \tag{4.10}$$

- **Power Law**

$$\log_a(x^k) = k\log_a(x). \tag{4.11}$$

- **Reciprocal Law**

$$\log_a\left(\frac{1}{x}\right) = \log_a(x^{-1}) = -\log_a(x). \tag{4.12}$$

Now suppose that $a > 0$, $a \neq 1$, $b > 0$ and $b \neq 1$, then, for $x > 0$, prove the following.

- **Change of Base**

$$\log_a(x) = \frac{\log_b(x)}{\log_b(a)}. \tag{4.13}$$

Solution:

The proofs follow on direct application of the definition of logarithms and laws of indices, shown in Example 4.14.

- **Multiplication Law**

  Let us begin with the product of two number x and y. We have, by the definition of logarithms, that

  $$xy = a^{\log_a(xy)}, \quad x = a^{\log_a(x)} \quad \text{and} \quad y = a^{\log_a(y)}.$$

  By rearranging the above and using the laws of indices for exponential functions, we have

  $$a^{\log_a(xy)} = xy$$
  $$= a^{\log_a(x)} \cdot a^{\log_a(y)}$$
  $$= a^{\log_a(x)+\log_a(y)}.$$

  It follows immediately that

  $$\log_a(xy) = \log_a(x) + \log_a(y).$$

- **Division Law**

  In an almost identical manner, we can consider $\frac{x}{y}$ and note that

  $$\frac{x}{y} = a^{\log_a\left(\frac{x}{y}\right)}.$$

  We then have

  $$a^{\log_a\left(\frac{x}{y}\right)} = \frac{x}{y}$$
  $$= a^{\log_a(x)}/a^{\log_a(y)}$$
  $$= a^{\log_a(x)-\log_a(y)}.$$

  It follows immediately that

  $$\log_a(x) - \log_a(y) = \log_a\left(\frac{x}{y}\right).$$

- **Power Law**
  For an x and exponent k, using the laws of indices gives:

  $$a^{\log_a(x^k)} = x^k$$
  $$= \left(a^{\log_a(x)}\right)^k$$
  $$= a^{k\log_a(x)}.$$

  Thus,

  $$\log_a(x^k) = k\log_a(x).$$

- **Reciprocal Law**
  The reciprocal law follows as a direct consequence of the power law with $k = -1$.
- **Change of Base**
  Firstly, we have

  $$a = b^{\log_b(a)}.$$

  Hence, for any $x > 0$, using the definition of logarithms and the laws of indices gives

  $$x = a^{\log_a(x)}$$
  $$= (b^{\log_b(a)})^{\log_a(x)}$$
  $$= b^{\log_b(a)\log_a(x)},$$
  $$\Rightarrow \quad \log_b(x) = \log_b(a)\log_a(x),$$
  $$\Rightarrow \quad \log_a(x) = \frac{\log_b(x)}{\log_b(a)}.$$

Example 4.17

Quadratic Equation Formula

Solutions to the general quadratic equation $ax^2 + bx + c = 0$ can be found using the formula

$$x = \frac{-b \pm \sqrt{b^2 - 4ac}}{2a}.$$

Solution:
The proof follows by completing the square on the general quadratic function ax^2+bx+c.

$$ax^2 + bx + c = 0,$$

$$\Rightarrow \qquad x^2 + \frac{b}{a}x + \frac{c}{a} = 0,$$

$$\Rightarrow \qquad x^2 + \frac{b}{a}x = -\frac{c}{a},$$

$$\Rightarrow \quad x^2 + \frac{b}{a}x + \left(\frac{b}{2a}\right)^2 = -\frac{c}{a} + \left(\frac{b}{2a}\right)^2,$$

$$\Rightarrow \qquad \left(x + \frac{b}{2a}\right)^2 = -\frac{c}{a} + \left(\frac{b}{2a}\right)^2,$$

$$\Rightarrow \qquad x + \frac{b}{2a} = \pm\sqrt{-\frac{c}{a} + \left(\frac{b}{2a}\right)^2},$$

$$\Rightarrow \qquad x = -\frac{b}{2a} \pm \sqrt{-\frac{c}{a} + \left(\frac{b}{2a}\right)^2},$$

$$\Rightarrow \qquad x = -\frac{b}{2a} \pm \sqrt{-\frac{c}{a} + \frac{b^2}{4a^2}},$$

$$\Rightarrow \qquad x = -\frac{b}{2a} \pm \sqrt{\frac{-4a^2c + b^2a}{4a^2}},$$

$$\Rightarrow \qquad x = -\frac{b}{2a} \pm \frac{1}{2a}\sqrt{-4ac + b^2},$$

$$\Rightarrow \qquad x = \frac{-b \pm \sqrt{b^2 - 4ac}}{2a}.$$

Example 4.18
Prove Pascal's identity (or Pascal's rule) which states that

$$\binom{n}{r-1} + \binom{n}{r} = \binom{n+1}{r}, \tag{4.14}$$

where $\binom{\cdot}{\cdot}$ is the binomial coefficient function.
Solution:
We can prove this identity by applying the definition of the binomial coefficient. Con-

sider the left hand side,

$$\binom{n}{r-1} + \binom{n}{r} = \frac{n!}{(r-1)!(n-(r-1))!} + \frac{n!}{r!(n-r)!},$$

$$= \frac{n!}{(r-1)!(n-r+1)!} + \frac{n!}{r!(n-r)!},$$

$$= n! \left[\frac{1}{(r-1)!(n-r+1)!} + \frac{1}{r!(n-r)!} \right],$$

$$= n! \left[\frac{r+(n-r+1)}{r!(n-r+1)!} \right],$$

$$= n! \left[\frac{n+1}{r!(n-r+1)!} \right],$$

$$= \frac{(n+1)!}{r!(n+1-r)!},$$

$$= \binom{n+1}{r}.$$

We now present from proofs of results from trigonometry and geometry.

Example 4.19

Show that the area A_T of a triangle ABC with sides of length a, b, and c is

$$A_T = \frac{1}{2} ab \sin(C).$$

Solution:

We consider two cases separately, one where the triangle has only acute angles, one where it has an obtuse angle.

Case 1:

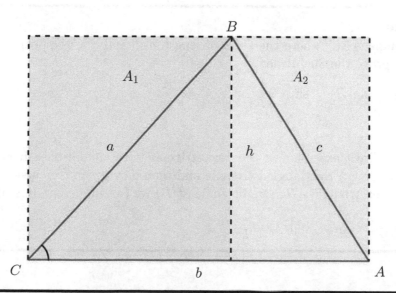

The figure above shows a triangle ABC and the rectangle which encloses it. ABC has all acute angles, or potentially one right angle.. Clearly the triangle can be split into two smaller right angled triangles, with areas $\frac{A_1}{2}$ and $\frac{A_2}{2}$. Hence, $A_T = \frac{A_1+A_2}{2}$. Now, $A_1 + A_2$ is the area of the enclosing rectangle and $A_1 + A_2 = bh$. h is found using trigonometry and is given by $h = a\sin(C)$. Thus, $A_T = \frac{1}{2}ab\sin(C)$.

Case 2:
Consider the figure below, which again shows a triangle ABC and a rectangle enclosing it. This time ABC has one obtuse angle.

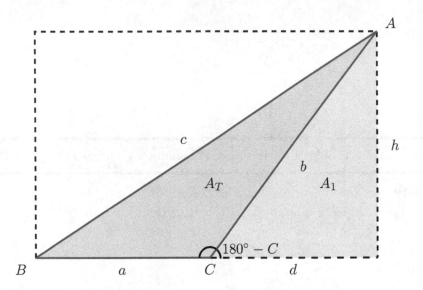

The area $A_T + A_1 = \frac{1}{2}(a + d)h$. Now, $A_1 = \frac{1}{2}dh$, hence $A_T = \frac{1}{2}ah$. It remains to find h. We have $h = b\sin(180° - C) = b\sin(C)$ using the periodic properties of sine. Thus, $A_T = ab\sin(C)$ in this case as well.

Example 4.20
For any triangle ABC, where the sides of length a, b and c are opposite A, B and C, respectively, prove the Sine Rule:

$$\frac{\sin(A)}{a} = \frac{\sin(B)}{b} = \frac{\sin(C)}{c}. \tag{4.15}$$

Solution:
This rule is a direct extension of the basic trigonometric definitions. It can be proved first by considering a right angled triangle and then moving on to a general triangle. In a right angled triangle ABC (right-angle at B) we can immediately state that:

$$\sin(A) = \frac{a}{b} \quad \text{and} \quad \sin(C) = \frac{c}{b}. \tag{4.16}$$

Rearranging these we find

$$\frac{\sin(A)}{a} = \frac{1}{b} = \frac{\sin(C)}{c}.$$

Since $\sin(90°) = 1$, we already have the sine rule relationship.

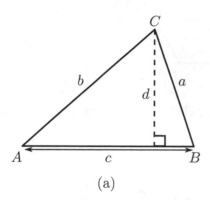 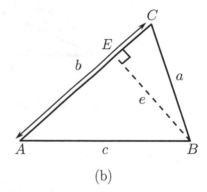

(a)   (b)

Figure 4.3: A triangle can be partitioned into two right-angled triangles in different ways.

We can now consider a more general triangle ABC shown in Figure 4.3. First it is partitioned into two right-angled triangles by dropping a perpendicular from vertex C to side AB, as in Figure 4.3(a). We use the ratios

$$\sin(A) = \frac{d}{b} \quad \text{and} \quad \sin(B) = \frac{d}{a}.$$

Rearranging gives

$$d = b\sin(A) = a\sin(B)$$
$$\Rightarrow \qquad \frac{\sin(A)}{a} = \frac{\sin(B)}{b}. \tag{4.17}$$

Alternatively, splitting into two right angles triangles by dropping a perpendicular from vertex B to side AC, as in Figure 4.3(b), leads us to the following equation

$$\frac{\sin(A)}{a} = \frac{\sin(C)}{c}. \tag{4.18}$$

The two equations (4.17) and (4.18) are combined to create the three part equation (4.15) which is the Sine rule.

If the triangle has an obtuse angle, the perpendicular will drop from the vertex to the opposite side *produced* (this means extended in a straight line). In this case we make use of the fact that $\sin(180° - \theta) = \sin(\theta)$.

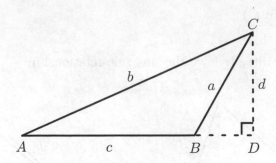

Figure 4.4: For a triangle with an obtuse angle, the line AB is produced to the point D.

Consider Figure 4.4, we have,

$$\sin(A) = \frac{d}{b} \qquad \text{and} \qquad \sin(180° - B) = \sin(B) = \frac{d}{a}.$$

These equations now rearrange as before.

Example 4.21

Consider a circle and suppose a tangent is drawn that touches the circle only at a point A, as shown in Figure 4.5. Prove that the tangent is perpendicular to the radius drawn from the centre of the circle to point A.

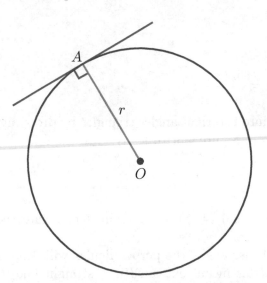

Figure 4.5: A tangent to a circle is perpendicular to the radius.

Solution:
In order to prove this, without loss of generality, we assume the circle is centred at the origin and therefore has equation $x^2 + y^2 = r^2$. Let point $A = (x_1, y_1)$ lie on the circle. We consider the case $x_1, y_1 \neq 0$ first.

Case 1:

In this case, the straight line which passes through the origin O and A has equation $y = m_1 x$, where $m_1 = y_1/x_1$. We now try to find the equation of the tangent to the circle at point A. This equation has general form $y = m_2 x + b$ and must intersect the circle *only* once at the point (x_1, y_1). Using the fact that the line intersects the circle at (x_1, y_1), we find that

$$y_1 = m_2 x_1 + b,$$
$$\Rightarrow \quad b = y_1 - m_2 x_1. \tag{4.19}$$

We also know that to find an intersection point of two curves/lines involves solving a set of simultaneous equations. Here, the two equations are

$$y = m_2 x + b, \quad \text{①}$$
$$r^2 = x^2 + y^2. \quad \text{②}$$

Substituting the expression for y from ① directly into ② gives

$$r^2 = x^2 + (m_2 x + b)^2.$$

This is a quadratic equation for x which can be rearranged into the usual form:

$$(1 + m_2^2)x^2 + 2bm_2 x + (b^2 + r^2) = 0.$$

As this is a quadratic equation, we expect there to be zero, one or two solutions, which can be found using the quadratic equation formula:

$$x = \frac{-2bm_2 \pm \sqrt{4b^2 m_2^2 - 4(1 + m_2^2)(b^2 + r^2)}}{2(1 + m_2^2)}.$$

Specifically, we are looking for the case where there is only one solution and hence, we know that the discriminant $4b^2 m_2^2 - 4(1 + m_2^2)(b^2 + r^2) = 0$ and

$$x_1 = -\frac{2bm_2}{2(1 + m_2^2)} = -\frac{bm_2}{(1 + m_2^2)}.$$

Now, using Equation 4.19, we have

$$x_1 = -\frac{(y_1 - m_2 x_1)m_2}{(1 + m_2^2)},$$
$$\Rightarrow \quad x_1(1 + m_2^2) = -(y_1 - m_2 x_1)m_2,$$
$$\Rightarrow \quad x_1 = -y_1 m_2,$$
$$\Rightarrow \quad m_2 = -\frac{x_1}{y_1}.$$

Now, the slope of the line passing through O and A is $m_1 = y_1/x_1$ and the slope of the tangent is $m_2 = -x_1/y_1$.

Case 2:

Let us suppose that $A = (0, r)$. In this case, the equation of the line passing through O and A is simply $x = 0$. Hence, a line perpendicular to this and passing through A would have equation $y = r$. We show that this is tangent to the circle at A, by showing it touches the circle just once. In order to do this, we solve the simultaneous equations

$$y = r, \quad ①$$
$$x^2 + y^2 = r^2, \quad ②.$$

Substituting ① into ② immediately yields $x^2 + r^2 = r^2$ and hence $x^2 = 0 \Rightarrow x = 0$. There is only one solution and reassuringly it is at the point $(0, r)$.

Case 3:

The above can be repeated at the point $A = (0, -r)$. By switching x and y, the same argument holds at the points $A = (-r, 0)$ and $A = (r, 0)$.

Proof Tip

In the above proof we used the phrase 'without loss of generality', sometimes this is shortened to just 'wlog'. It essentially means that only one or a number of specific cases need to be considered to prove the result. In the above, clearly not all circles will have centre at the origin, but a circle with radius r centred at the origin is really no different to a circle of radius r centred anywhere else - the circle is invariant under translations. In fact, as circles are also invariant under rotations, it would suffice to consider only one of Cases 2 or 3, but Case 1 is perhaps more instructive and gives added practice in algebraic manipulation.

In general, there is no rule for when 'without loss of generality' can be used, it just depends on the situation.

Proof Tip

The method shown above, for proving that the tangent and radius are perpendicular, is very long and algebraically heavy. Differentiation would have enabled us to find the equations of the tangent much more simply.

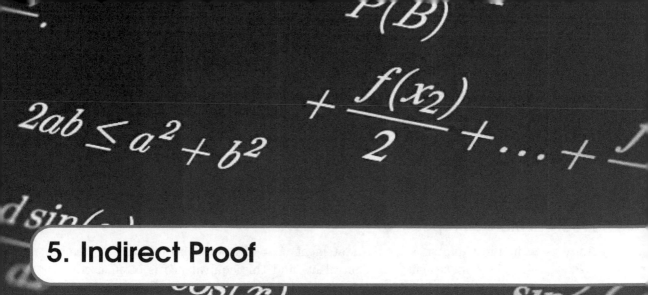

$$2ab \leq a^2 + b^2$$

$$+ \frac{f(x_2)}{2} + \ldots +$$

5. Indirect Proof

While methods of direct proof are appropriate in many cases, there are several statements that cannot be proved directly.

Figure 5.1: Euclid of Alexandria, circa 300 BCE.

The Greek mathematician Euclid of Alexandria is often referred to as the "father of geometry", and is arguably one of the most influential figures in the history of mathematics. Amongst his many contributions to mathematics, Euclid provided a proof by contradiction that there are infinitely many primes, a key theorem in number theory. We shall study this proof later in this chapter.

Prime numbers have been studied at length for over 2000 years since Euclid provided this proof. Prime numbers are at the heart of number theory's applied counterpart, *cryptograpy*. When we enter personal information or make purchases on "secure" web pages, prime numbers are used as *keys*, which an algorithm uses to *encrypts* our information. The web page is secure if only the page's owner can decrypt our encoded information using the correct key. In the event that our information is intercepted, or the web page hacked by unscrupulous parties, the security of our information depends on theorems associated with prime numbers.

In this chapter we consider two methods of indirect proof, namely "proof via the contrapositive" and "proof by contradiction".

5.1 Proof via the Contrapositive

As we saw in the Chapter 3 for the statement $A \Rightarrow B$, the contrapositive statement is $\neg B \Rightarrow \neg A$. These statements are equivalent, and this equivalence is often useful when trying to prove some results. It may be that the contrapositive of what we are trying to show is actually easier to prove than the original statement. Once the contrapositive has been proved we can then infer the truth of the original statement.

Example 5.1

Let $x \in \mathbb{Z}$. Prove that if $3x + 7$ is even then x is odd.

Solution:

The contrapositive of this statement is "If x is not odd then $3x + 7$ is not even". This contrapositive could be equivalently written "If x is even then $3x + 7$ is odd" as integers are either even or odd.

To prove this contrapositive let $x = 2m$ for some integer m, then,

$$
\begin{aligned}
3x + 7 &= 3(2m) + 7 \\
&= 6m + 6 + 1 \\
&= 2(3m + 3) + 1,
\end{aligned}
$$

which is odd. Therefore, we have proved the contrapositive statement and so can now infer the truth of our original statement.

Proof Tip

We could have proved this directly in the following way:

Suppose that $3x + 7$ is even, and so we can write $3x + 7 = 2m$ for some integer m. Subtracting $2x + 7$ from both sides we have,

$$
\begin{aligned}
x &= 2m - 2x - 7 \\
&= 2(m - x) - 8 + 1 \\
&= 2(m - x - 4) + 1.
\end{aligned}
$$

Consequently x is of the form $x = 2n + 1$, for some integer n (in fact, by construction, $n = m - x - 4$) and is therefore odd.

This direct proof is about the same length of the proof of the contrapositive, but it could be argued that the proof by the contrapositive flows more smoothly.

Example 5.2

Prove that for $x, y \in \mathbb{Z}$ with $x + y$ even, then x and y have the same parity (*i.e.* both integers x,y are either even or odd).

Solution:

The contrapositive of this result states that "If x and y are two integers of opposite parity, then their sum must be odd".

Assuming, without loss of generality, that x is even and y is odd, we may write $x = 2m$ and $y = 2n + 1$ for some integers m and n. Now,

$$x + y = 2m + (2n + 1)$$
$$= 2(m + n) + 1,$$

which by definition must be an odd integer.
Therefore, we have proved the contrapositive statement and so can now infer the truth of our original statement.

Proof Tip

Note that in the above, even though we are proving the result with an indirect proof we have used a direct proof to prove the contrapositive of our original statement.

Example 5.3

Prove the following statement for $x \in \mathbb{Z}$:

> If x^2 is even then x is even.   (5.1)

Solution:

The contrapositive of statement (5.1) is "If x is odd then x^2 is also odd".
To prove the contrapositive we let $x = 2m + 1$, for some integer m. Then,

$$x^2 = (2m + 1)^2$$
$$= 4m^2 + 4m + 1$$
$$= 2(2m^2 + 2m) + 1,$$

which is odd. Therefore, we have proved the contrapositive statement and so can now infer the truth of our original statement.

Example 5.4

Prove that every prime number is either odd or two using a contrapositive argument.

Solution:

We prove directly the contrapositive statement "n is neither odd nor two then n is not prime". Let $n > 0$, $n \in \mathbb{Z}$. If n is neither odd nor two then $n = 2k$ for some integer k which is not 1. Hence n is not prime.

Exercise 5.1

Q1. Let $x \in \mathbb{Z}$. Prove that if $x^2 - 8x + 7$ is even then x is odd.

Q2. Prove that for $x, y \in \mathbb{Z}$ where xy is even, then at least one of x or y must be even.

5.2 Proof by Contradiction

Proof by contradiction works in the following way. Consider a statement A, for which we may not have a direct proof. We assume that A is *false*. If, by performing a series of logical steps (or algebraic manipulations), we arrive at an inconsistency, we conclude that A is true. This is because the contradiction was reached under the assumption that A was false.

Example 5.5

Prove that the sum of two odd numbers is always even.

Solution:

Suppose, for contradiction, that there exist two odd numbers whose sum is odd. Then, for positive integers a and b, and positive odd integer c, we have

$$
\begin{aligned}
& (2a + 1) + (2b + 1) = c, \\
\Rightarrow \quad & 2(a + b + 1) = c, \\
\Rightarrow \quad & a + b + 1 = \frac{c}{2}.
\end{aligned}
$$

Since we assumed that c is odd, it is not divisible by 2, so $\frac{c}{2}$ *is* not an integer. However, $a + b + 1$ *is* an integer, and we have arrived at a contradiction. Hence, there are no such odd integers whose sum is odd and we conclude that the sum of any two odd numbers is even.

Remark

Although the proof in Example 5.5 is correct, it is neater and more efficient to prove directly that the sum of two odd numbers is even.

Example 5.6

The *arithmetic-geometric mean inequality* (seen in the introduction to this book) states that, for two non-negative numbers a and b,

$$\frac{1}{2}(a + b) \geq \sqrt{ab}. \tag{5.2}$$

Prove (5.2) by contradiction.

Solution:

We assume that (5.2) is false. Under this assumption,

$$\frac{1}{2}(a + b) < \sqrt{ab}.$$

By performing some algebraic manipulation, we find

$$\frac{1}{2}(a + b) < \sqrt{ab},$$
$$\Rightarrow \quad \frac{1}{4}(a + b)^2 < ab,$$
$$\Rightarrow \quad (a + b)^2 < 4ab,$$
$$\Rightarrow \quad a^2 + 2ab + b^2 < 4ab,$$
$$\Rightarrow \quad a^2 - 2ab + b^2 < 0,$$
$$\Rightarrow \quad (a - b)^2 < 0.$$

We have arrived at a contradiction, since the square of any number is non-negative. Since our only assumption was that (5.2) is false, we conclude that (5.2) is true.

In the following example we give the proof attributed to Euclid as mentioned in the introduction to this chapter. For this, we will need the the intermediate result.

Theorem 5.1 — Existence of a Prime Factor

Every integer $M \geq 2$ has a prime factor.

Proof:

If M is itself prime, then this completes the proof. If M is not prime, then M has a factor M_1 with $1 < M_1 < M$. Any factor of M_1 is a factor of M. Now, either M_1 is prime, or M_1 has a factor M_2 with $1 < M_2 < M_1$. Repeating this argument finitely many times, we obtain a prime factor of M.

Example 5.7

Prove the statement *"there are infinitely many prime numbers"*.

Solution:

This famous proof is attributed to Euclid, who devised it over 2000 years ago. We begin by assuming that the statement is false, then there are only finitely many primes, say n of them. Write them down in increasing order such that

$$p_1 = 2 < p_2 = 3 < p_3 < \cdots < p_n.$$

Now let

$$M = p_1 p_2 \ldots p_n + 1.$$

Since $M \geq 2$, it has a prime factor, which must be one of the $p_1, \ldots, p_n$, say p_j $(j < n)$. This is a contradiction, since dividing M by p_j leaves remainder 1. Our assumption that there are only finitely many primes has led to a contradiction, so we conclude that there are infinitely many primes.

Proof Tip

It is a common misconception to claim that the proof in Example 5.7 shows that prime numbers can be generated by multiplying together the first n primes and adding 1.

This claim is wrong; this can be shown by finding a counterexample.
The first few values of n do indeed produce prime numbers:

$$
\begin{aligned}
n = 1: \quad & 2 + 1 = 3, \\
n = 2: \quad & 2 \times 3 + 1 = 7, \\
n = 3: \quad & 2 \times 3 \times 5 + 1 = 31, \\
n = 4: \quad & 2 \times 3 \times 5 \times 7 + 1 = 211, \\
n = 5: \quad & 2 \times 3 \times 5 \times 7 \times 11 + 1 = 2311.
\end{aligned}
$$

However, when $n = 6$,

$$
\begin{aligned}
2 \times 3 \times 5 \times 7 \times 11 \times 13 + 1 &= 30\,301 \\
&= 59 \times 509.
\end{aligned}
$$

Hence, $30\,301$ has a prime factor that is in fact greater than p_6.

The remark that follows is a worthy discussion regarding the implications to cryptography, of the claim that a formula can be used to generate primes.

Remark

Secure information, such as credit card details that are entered to make online transactions must be encrypted for security. One of the most common encryption schemes is the RSA algorithm, designed by the mathematicians Ron Rivest, Adi Shamir and Leonard Adleman. The algorithm uses a *public key*, which is information that is available to everybody, and a *private key*, which only the decoding party (typically the merchant) has.
The details of the complete algorithm are not important here, and require knowledge of more advanced mathematics such as modulo arithmetic. One of the steps in the RSA algorithm is the generation of a positive integer N, which is the product of two very

large (usually hundreds of digits long) prime numbers p and q. The number N forms part of the public key, while p and q are only known to the merchant. If the public key, including the value of N, is intercepted, the only way to decrypt the information is to find the prime factors p and q. For sufficiently large N, however, it may take many months to find the prime factors of N, even using a supercomputer. It is important that p and q are prime numbers since there are no (current) methods that can factor a number into its prime factors efficiently. If it were possible to guarantee the generation of all prime numbers with some formula, then we would increase the efficiency of finding the prime factors of N. We could simply generate prime numbers and check systematically whether they are factors of N and hence Internet security would be at risk.

Example 5.8

In Chapter 4 we proved directly, for positive integer n, that

$$n \text{ even} \Rightarrow n^2 \text{ even,}$$
$$n \text{ odd} \Rightarrow n^2 \text{ odd.}$$

Prove by contradiction that the reverse implications

$$n^2 \text{ even} \Rightarrow n \text{ even,}$$
$$n^2 \text{ odd} \Rightarrow n \text{ odd}$$

are also true.

Solution:
We begin with the statement "n^2 even $\Rightarrow n$ even".
We are given that n^2 is even. Suppose, for contradiction, that n is odd. Then there exists a positive integer k such that $n = 2k + 1$. Then

$$n^2 = (2k + 1)^2 = 4k^2 + 4k + 1 = 2(2k^2 + 2k) + 1.$$

Since $2k^2 + 2k$ is a positive integer, n^2 is odd, which contradicts our premise that n^2 is even. Since we have arrived at a contradiction by assuming that n is odd if n^2 is even, we conclude that n^2 even $\Rightarrow n$ even.

The proof of the statement "n^2 odd $\Rightarrow n$ odd" is almost identical. We are given that n^2 is odd. Suppose, for contradiction, that n is even. Then there exists a positive integer k such that $n = 2k$. Then

$$n^2 = 4k^2 = 2(2k^2).$$

Since $2k^2$ is a positive integer, n^2 is even, which contradicts our premise that n^2 is odd. Since we have arrived at a contradiction by assuming that n is even if n^2 is odd, we conclude that n^2 odd $\Rightarrow n$ odd.

Exercise 5.2

Q1. Prove by contradiction that the sum of two even numbers is even.

Q2. Prove that there is no maximum even number.

Q3. Prove that there exist no integers a and b for which $15a + 3b = 1$.

Q4. Let n be a positive integer. Prove that if $n^3 + 5$ is odd, then n is even.

Q5. Prove that, for every real number x, where $0 \leq x \leq \pi/2$, we have $\sin(x) + \cos(x) \geq 1$.

Q6. (a) Let a, b and c be positive integers. Prove directly that if a is a factor of b and b is a factor of c, then a is a factor of c.

   (b) Let $n \geq 2$ be a positive integer. We know that n has at least two positive factors, namely 1 and n. Prove that the *second smallest* positive factor of n must be a prime number.

   Hint: Use the result in (a).

Q7. Prove that the sum of two positive integers is positive.

Q8. Prove that there is no least positive rational number.

Q9. Suppose a sequence (a_n) is defined by the rules

$$a_1 = 4; \quad a_2 = 26; \quad a_n = 4a_{n-1} + 5a_{n-2} \text{ for } n \geq 3.$$

Prove that $a_n = 5^n + (-1)^n$ for every positive integer n.

Hint: Let k be the least positive integer such that $a_k \neq 5^k + (-1)^k$ and seek a contradiction. In this type of proof, we refer to k as the "minimal criminal".

Q10. Prove that every positive integer $n \geq 2$ is a product of prime numbers.

Q11. Prove that there exist no integers a and b such that $a^2 - 4b = 2$.

Q12. Prove that there are no positive integer solutions to the equation $x^2 - y^2 = 1$.

Q13. Prove that, for all $n \in \mathbb{N}$,

$$\frac{n}{n+1} < 1.$$

Q14. Prove that, for $n \in \mathbb{Z}$, if $3n + 2$ is even, then n is even.

Q15. Prove that, for $a \in \mathbb{Z}$, If $a^2 - 2a + 7$ is even, then a is odd.

Q16. Prove that the equation $2x^3 + 6x + 1 = 0$ has no integer solutions.

Q17. Prove that for every positive $x \in \mathbb{R}$,

$$x + \frac{1}{x} \geq 2.$$

Q18. Prove that for every positive $x \in \mathbb{R}$,

$$\frac{x}{x+1} < \frac{x+1}{x+2}.$$

.1 Proof by Contradiction for Irrationality

Proof by contradiction allows us to prove that some numbers are irrational.

Definition 5.2 — Rational and Irrational Numbers

The number n is a *rational* number if there exists an integer a, and a positive integer b such that

$$n = \frac{a}{b}.$$

If n cannot be represented in this way, then it is *irrational*. Examples of irrational numbers are π, e and $\sqrt{2}$.

We first prove some basic properties of irrational numbers using direct proof.

Example 5.9

Let x and y be rational numbers. Prove directly that xy, $x + y$ and $x - y$ are also rational. Prove that, if $y \neq 0$, $\dfrac{x}{y}$ is also rational.

Solution:

Given that x and y are rational, there exist integers a_1 and a_2, and positive integers b_1 and b_2 such that

$$x = \frac{a_1}{b_1} \text{ and } y = \frac{a_2}{b_2}.$$

Then

$$xy = \frac{a_1 a_2}{b_1 b_2}.$$

Since $a_1 a_2$ is an integer and $b_1 b_2$ is a positive integer, xy satisfies the definition of a rational number.

Also,

$$x \pm y = \frac{a_1}{b_1} \pm \frac{a_2}{b_2} = \frac{a_1 b_2 \pm b_1 a_2}{b_1 b_2}.$$

Since $a_1 b_2 \pm b_1 a_2$ is an integer and $b_1 b_2$ is a positive integer, both $x + y$ and $x - y$ are rational.

Finally, for $y \neq 0$ we have $a_2 \neq 0$, and then

$$\frac{x}{y} = \frac{a_1/b_1}{a_2/b_2} = \frac{a_1 b_2}{a_2 b_1}.$$

Clearly, $a_1 b_2$ is an integer. Also, $a_2 b_1$ is a non-zero integer. If $a_2 b_1 > 0$, then $a_2 b_1$ is a positive integer and x/y is rational. Otherwise, $-a_2 b_1 > 0$ and

$$\frac{x}{y} = \frac{-a_1 b_2}{-a_2 b_1}$$

is rational.

Example 5.10

Prove that $\log_2(3)$ is irrational.

Solution:

Suppose, for contradiction, that $\log_2(3)$ is rational. Then there exist $a \in \mathbb{Z}$ and $b \in \mathbb{N}$ such that

$$\log_2(3) = \frac{a}{b}$$
$$\Rightarrow \quad b\log_2(3) = a$$
$$\Rightarrow \quad 2^a = 3^b.$$

Since 2^a is even for all a and 3^b is odd for all b, the final equality is impossible. Hence, we have arrived at a contradiction and conclude that $\log_2(3)$ is irrational.

Example 5.11

Prove that the negative of any irrational number is also irrational.

Solution:

Suppose, for contradiction, that there exists an irrational number n such that $-n$ is rational. Under this assumption, there exist integers a and $b > 0$ such that

$$-n = \frac{a}{b}.$$

Dividing both sides by -1,

$$n = -\frac{a}{b} = \frac{-a}{b}.$$

Since a is an integer, $-a$ is also an integer and so n is rational. We have arrived at a contradiction, since we initially assumed that n is irrational. We conclude that the negative of any irrational number is irrational.

Throughout history the irrationality of π has fascinated many mathematicians and several proofs have been provided that π is irrational. The first of which was published in the 18th century by Johann Heinrich Lambert. In the 19th century others, such as Charles Hermite, provided alternative proofs by contradiction of the irrationality of π. Another fundamental constant, e, is also irrational. Joseph Fourier provided a proof by contradiction in the 19th century of the irrationality of e. The details of these proofs are presented at the end of the chapter.

Before this however we consider the proof of the irrationality of $\sqrt{2}$. This is often one of the first indirect proofs students of mathematics see and there are many ways to prove this; the most popular are considered in the example below.

Example 5.12
Prove that $\sqrt{2}$ is irrational.

Solution:
There are several proofs of the irrationality of $\sqrt{2}$. Some of them are as follows.

1. Recall that a number n is *rational* if it can be expressed as $n = \dfrac{a}{b}$, where a and b are integers and $b > 0$. Otherwise, n is *irrational*. Suppose, for contradiction, that $\sqrt{2}$ is rational. Under this assumption, there exist positive integers a and b such that

$$\sqrt{2} = \frac{a}{b}.$$

In particular, we insist that the fraction $\dfrac{a}{b}$ is in its lowest terms. Equivalently, we say that the highest common factor of a and b is 1.
Performing some algebraic manipulation, we find

$$\sqrt{2} = \frac{a}{b},$$
$$\Rightarrow \quad 2 = \frac{a^2}{b^2},$$
$$\Rightarrow \quad 2b^2 = a^2.$$

Hence, a^2 is even. By the result in Example 5.8, a is also even, so can be expressed as $a = 2k$, where k is a positive integer. Performing further algebraic manipulation, we find

$$\Rightarrow \quad 2b^2 = 4k^2,$$
$$\Rightarrow \quad b^2 = 2k^2.$$

Hence, b^2 is also even and thus so is b. Since both a and b are even, the highest common factor of a and b is at least $2 > 1$. This contradicts our assumption that $\dfrac{a}{b}$ was written in its lowest terms. Since the assumption that $\sqrt{2}$ is rational has led to a contradiction, we conclude that $\sqrt{2}$ is irrational.

2. The *Fundamental Theorem of Arithmetic* states that every integer is uniquely (up to the ordering of the factors) representable as a product of primes. Assume, as above, that $\sqrt{2}$ is rational, which means that there exist positive integers a and b such that

$$2b^2 = a^2.$$

Factoring a^2 into a product of primes, a^2 has the same prime factors as a, each taken twice. Thus, a^2 has an even number of prime factors, as does b^2. Since 2 is prime, $2b^2$ has an odd number of prime factors. Combined with $2b^2 = a^2$, we have arrived at a contradiction and so conclude that $\sqrt{2}$ is irrational.

3. Assume, for contradiction, that $\sqrt{2}$ is rational. Furthermore, assume that there exist positive integers a and b such that $\sqrt{2} = \dfrac{a}{b}$, where b is as small as possible.

Then $a > b$ and also $b > a - b$, so we have

$$\frac{2b - a}{a - b} = \frac{2 - a/b}{a/b - 1}$$

$$= \frac{2 - \sqrt{2}}{\sqrt{2} - 1}$$

$$= (2 - \sqrt{2})(\sqrt{2} + 1)$$

$$= \sqrt{2}.$$

Hence, $\sqrt{2}$ can be represented as $\dfrac{2b - a}{a - b}$, but $b > a - b$, so this contradicts the minimality of b. Thus, we conclude that $\sqrt{2}$ is irrational.

Proof Tip

The first proof is the "standard" proof of the irrationality of $\sqrt{2}$. The steps are the most obvious of the three, since it requires the least background knowledge, such as the Fundamental Theorem of Arithmetic. This proof can also be generalised to prove the irrationality of other numbers (see Exercise 5.3).

The first half of proof 2 follows the same algebraic manipulation of proof 1, then is an effective application of the Fundamental Theorem of Arithmetic, though students would already need to know about the theorem. Since this proof only involved comparing the number of prime factors on either side of the equation that results from the same algebraic manipulation as in proof 1, this proof can be generalised to prove the irrationality of other numbers.

Proof 3 involves calculations which are, perhaps, less obvious. Another issue with this proof is that there is no obvious way to generalise it to prove the irrationality of other numbers.

Overall, in our opinion, proof 1 is the best line of argument to use. Only knowledge of relatively basic knowledge of mathematics is required to follow and reproduce this proof, and it can be easily generalised to prove the irrationality of other numbers.

Exercise 5.3

Q1. Let x and y be rational numbers with $y \neq 0$. Prove that $x + y\sqrt{2}$ is irrational.

Q2. Prove that the sum of a rational number and an irrational number is irrational.

Q3. (a) Let n be a positive integer. Prove that if n^2 is divisible by 3, then n is divisible by 3.
  (b) Use the result in (a) to prove that $\sqrt{3}$ is irrational.

Q4. (a) Prove that $\sqrt{6}$ is irrational.
  (b) Can we adapt the proof in (a) to prove that $\sqrt{8}$ is irrational.
  (c) What happens if we try to adapt the proof in (a) to prove that $\sqrt{9}$ is irrational?

Q5. (a) Prove that, for all positive integers n, that if n^3 is even, then n is even.
  (b) Use the result in (a) to prove that $2^{1/3}$ is irrational.

Q6. Let a, b and c be positive real numbers. Prove that if $ab = c$, then $a \leq \sqrt{c}$ or $b \leq \sqrt{c}$.

Q7. Let a, b and c be odd integers. Prove that the quadratic equation

$$ax^2 + bx + c = 0$$

has no rational solutions.

Q8. Prove that every non-zero rational number can be expressed as the product of two irrational numbers.

Q9. Prove or disprove the statement that, if a and b are rational, then a^b is also rational.

Q10. (a) Let $x \in \mathbb{R}$ be positive. Prove that if x is irrational, then $x^{1/6}$ is also irrational.
(b) Is it true that if $x^{1/6}$ is irrational, then x is irrational?

Q11. Fermat's method of infinite descent is based on the fact that there exists a smallest positive integer, namely 1. If some premise leads to infinitely many integer solutions, where each successive solution is smaller than the last, then this is a contradiction.

Use Fermat's method of infinite descent to prove that if k is a positive integer and $\sqrt{k}$ is not an integer, then $\sqrt{k}$ is irrational. Follow these steps:

- Assume that $\sqrt{k}$ is rational, so can be expressed as $\dfrac{a}{b}$, where a and b are positive integers with highest common factor 1.
- Let n be the largest integer less than $\sqrt{k}$, and express $\dfrac{a}{b}$ as

$$\frac{a}{b} = \frac{a\left(\sqrt{k} - n\right)}{b\left(\sqrt{k} - n\right)}.$$

- Perform some algebraic manipulations to arrive at a contradiction.

.2 Irrationality of Mathematical Constants

Here we present proofs of the irrationality of e and π.

Example 5.13

Prove that e (Euler's number) is irrational.

Solution:

We begin with Bernoulli's definition e, namely

$$e = \lim_{n \to \infty} \left(1 + \frac{1}{n}\right)^n.$$

It can be shown that

$$\lim_{n \to \infty} \left(1 + \frac{1}{n}\right)^n = \sum_{k=0}^{\infty} \frac{1}{k!}.$$

Hence, we can write

$$e = \frac{1}{0!} + \frac{1}{1!} + \frac{1}{2!} + \ldots + \frac{1}{n!} + R_n, \quad \text{where} \quad 0 < R_n < \frac{3}{(n+1)!}$$

Suppose, for contradiction, that e were rational so that there exist positive integers a and b such that $e = \frac{a}{b}$. Choose $n > b$ and also $n > 3$. Then

$$\frac{a}{b} = 1 + 1 + \frac{1}{2!} + \ldots + \frac{1}{n!} + R_n$$

Multiplying both sides by $n!$,

$$\frac{n!a}{b} = n! + n! + \frac{n!}{2!} + \ldots + \frac{n!}{n!} + n!R_n.$$

Since $n > b$, $n!$ is divisible by b for all n. Hence, every term in the equation above, apart from $n!R_n$ is clearly an integer. For the equation to hold, $n!R_n$ must be an integer too. However,

$$0 < R_n < \frac{3}{(n+1)!} \quad \Rightarrow \quad 0 < n!R_n < \frac{3}{n+1}.$$

Since we chose $n > 3$,

$$\frac{3}{n+1} < \frac{3}{4} < 1,$$

which is impossible for integer $n!R_n$. We have arrived at a contradiction, so our assumption that e is rational is false, and so we conclude that e is irrational.

Example 5.14
Prove that π is irrational.

Solution:
Although it is a well-known fact that π is an irrational number, the proof is surprisingly difficult. Before we begin, we must state and prove two observations that facilitate the proof of the irrationality of π.

Observation 1:
Consider the function

$$f_n(x) = \frac{x^n(1-x)^n}{n!},$$

which satisfies

$$0 < f_n(x) < \frac{1}{n!} \quad \text{for} \quad 0 < x < 1.$$

Multiplying out $x^n(1-x)^n$, the lowest power of x is n and the highest power is $2n$.

Thus, we can write $f_n(x)$ in the form

$$f_n(x) = \frac{1}{n!} \sum_{k=n}^{2n} c_k x^k,$$

where the coefficients c_k are integers. It is clear from this expression that

$$f_n^{(k)}(0) = 0 \quad \text{for} \quad k < n \text{ or } k > 2n.$$

Furthermore,

$$f_n^{(n)}(x) = \frac{1}{n!}[n!c_n + \text{terms involving } x],$$

$$\text{(All other terms differentiate to zero.)}$$

$$f_n^{(n+1)}(x) = \frac{1}{n!}[(n+1)!c_{n+1} + \text{terms involving } x],$$

$$\vdots$$

$$f_n^{(2n)}(x) = \frac{1}{n!}[(2n)!c_{2n}].$$

Setting $x = 0$,

$$f_n^{(n)}(0) = c_n,$$

$$f_n^{(n+1)}(0) = (n+1)c_{n+1},$$

$$\vdots$$

$$f_n^{(2n)}(0) = (2n)(2n-1)\cdots(n+1)c_{2n}.$$

Since n and all the c_k are integers, $f_n^{(k)}(0)$ is an integer for all k.

The relation

$$f_n(x) = f_n(1-x),$$

along with the chain rule, implies that

$$f_n^{(k)}(x) = (-1)^k f_n^{(k)}(1-x).$$

Hence, $f_n^{(k)}(1)$ is an integer for all k.

Observation 2:
Let $a \in \mathbb{R}^+$ and $\varepsilon > 0$. Then, for sufficiently large n, we have

$$\frac{a^n}{n!} < \varepsilon.$$

To prove this, we note that if $n \geq 2a$, then

$$\frac{a^{n+1}}{(n+1)!} = \frac{a}{n+1} \cdot \frac{a^n}{n!} < \frac{1}{2} \cdot \frac{a^n}{n!}.$$

Let $n_0 \in \mathbb{N}$ with $n_0 \geq 2a$. Then, whatever value

$$\frac{a^{n_0}}{n_0!}$$

may have, the succeeding values satisfy

$$\frac{a^{n_0+1}}{(n_0+1)!} < \frac{1}{2} \cdot \frac{a^{n_0}}{n_0!},$$

$$\frac{a^{n_0+2}}{(n_0+2)!} < \frac{1}{2} \cdot \frac{1}{2} \cdot \frac{a^{n_0}}{n_0!},$$

$$\vdots$$

$$\frac{a^{n_0+k}}{(n_0+k)!} < \frac{1}{2^k} \cdot \frac{a^{n_0}}{n_0!}.$$

If k is large enough such that

$$\frac{a^{n_0}}{n_0! \varepsilon} < 2^k,$$

then

$$\frac{a^{n_0+k}}{(n_0+k)!} < \varepsilon,$$

which finally gives us the desired result:

$$\frac{a^n}{n!} < \varepsilon$$

for sufficiently large n.

Proof:

We are now in a position to proceed with the proof that π is irrational. In fact, we prove that π^2 is irrational. The irrationality of π^2 implies the irrationality of π. If π were rational, then π^2 certainly would be.

Suppose, for contradiction, that π^2 were rational, so that there exist positive integers a and b such that

$$\pi^2 = \frac{a}{b}.$$

Let

$$G(x) = b^n \left[\pi^{2n} f_n(x) - \pi^{2n-2} f_n''(x) + \pi^{2n-4} f_n^{(4)}(x) + \ldots + (-1)^n f_n^{(2n)}(x) \right].$$

Each of the factors

$$b^n \pi^{2n-2k} = b^n (\pi^2)^{n-k} = b^n \left(\frac{a}{b}\right)^{n-k} = a^{n-k} b^k$$

is an integer. Using observation 1, $f_n^{(k)}(0)$ and $f_n^{(k)}(1)$ are integers for all k. Hence, $G(0)$ and $G(1)$ are integers. Differentiating $G(x)$ twice yields

$$G''(x) = b^n \left[\pi^{2n} f_n''(x) - \pi^{2n-2} f_n^{(4)}(x) + \ldots + (-1)^n f_n^{(2n+2)}(x) \right].$$

The final term, $(-1)^n f_n^{(2n+2)}(x)$, is zero. Thus,

$$G''(x) + \pi^2 G(x) = b^n \pi^{2n+2} f_n(x) = \pi^2 a^n f_n(x).$$

Now let

$$H(x) = G'(x) \sin(\pi x) - \pi G(x) \cos(\pi x).$$

Then

$$\begin{aligned} H'(x) &= \pi G'(x) \cos(\pi x) + G''(x) \sin(\pi x) - \pi G'(x) \cos(\pi x) + \pi^2 G(x) \sin(\pi x), \\ &= \left[G''(x) + \pi^2 G(x) \right] \sin(\pi x), \\ &= \pi^2 a^n f_n(x) \sin(\pi x). \end{aligned}$$

By the Fundamental Theorem of Calculus,

$$\begin{aligned} \pi^2 \int_0^1 a^n f_n(x) \sin(\pi x) \, \mathrm{d}x &= \int_0^1 H'(x) \, \mathrm{d}x, \\ &= H(1) - H(0), \\ &= G'(1) \sin \pi - \pi G(1) \cos \pi - G'(0) \sin 0 + \pi G(0) \cos 0, \\ &= \pi \left[G(1) + G(0) \right]. \end{aligned}$$

Thus, since $G(0)$ and $G(1)$ are integers,

$$\pi \int_0^1 a^n f_n(x) \sin(\pi x) \, \mathrm{d}x = G(1) + G(0)$$

is an integer.

Recall that the function $f_n(x)$ has the property

$$0 < f_n(x) < \frac{1}{n!} \quad \text{for} \quad 0 < x < 1,$$

so

$$0 < \pi a^n f_n(x) \sin(\pi x) < \frac{\pi a^n}{n!} \quad \text{for} \quad 0 < x < 1.$$

Consequently,

$$0 < \pi \int_0^1 a^n f_n(x) \sin(\pi x) \, \mathrm{d}x < \frac{\pi a^n}{n!}.$$

Note that this reasoning is independent of n. By observation 2, if n is sufficiently large, then

$$0 < \pi \int_0^1 a^n f_n(x) \sin(\pi x) \, \mathrm{d}x < \frac{\pi a^n}{n!} < 1.$$

Finally, we arrive at a contradiction. By assuming π^2 were rational, we deduced that this integral must be an integer, but there is no integer between 0 and 1. Hence, our assumption that π^2 is rational is incorrect and we conclude that π^2, and hence π, is irrational.

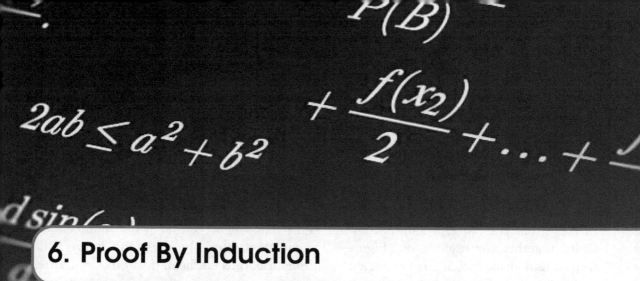

6. Proof By Induction

Proof by mathematical induction is a form of direct proof that is commonly used to prove statements that are indexed by a natural number n.

The Italian mathematician Francesco Maurolico (1494-1575) is often credited with the first published use of mathematical induction (albeit in an informal sense) in his 1575 book "Arithmeticorum libri duo".

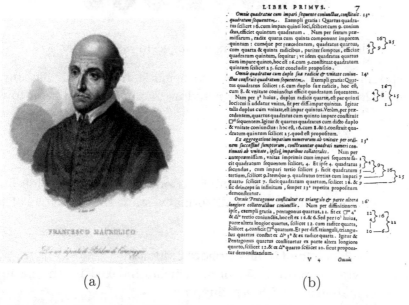

(a)   (b)

Figure 6.1: Francesco Maurolico (a) and page 7 of his "Arithmeticorum libri duo" (b)

On page 7 of this text he explores the addition of consecutive odd numbers and shows that this always results in a square number (the examples he used can be seen in the right margin on Figure 6.1 (b)).

Exploring a few of these sums certainly suggests this result to be true:

$$1 + 3 = 4,$$
$$1 + 3 + 5 = 9,$$
$$1 + 3 + 5 + 7 = 16,$$
$$1 + 3 + 5 + 7 + 9 = 25,$$
$$1 + 3 + 5 + 7 + 9 + 11 = 36.$$

In fact, from the patterns shown above it seems that when the first n odd numbers are added together the result is n^2. Figure 6.2 shows this pictorially; often this image is described as a visual "proof". However, a more formal, rigorous mathematical proof is desired.

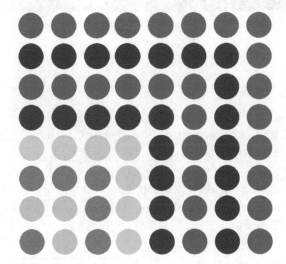

Figure 6.2: Odd numbers arranged in a square lattice.

To make the statement of this result more mathematical, we recall that any odd number can be written as $2r - 1$ for some integer r and use summation notation to write

$$\sum_{r=1}^{n} (2r - 1) = n^2.$$

We shall formally prove this statement about the odd numbers in Section 6.1.

Before attempting any examples we shall formally state the principle of mathematical induction.

> **Theorem 6.1 — The Principle of Mathematical Induction**
> Suppose that we wish to prove a set of statements indexed by the natural numbers.
> Denote this set of statements by $P(n)$. Suppose,
> - $P(1)$ is true;
> - For any $n = k$ if $P(k)$ is true then so is $P(k+1)$;
>
> then $P(n)$ is true for all $n \in \mathbb{N}$.

Proof:
Let F denote the subset of $\mathbb{N}$ consisting of all n for which $P(n)$ is false. If we can show that $F = \emptyset$ then we are done.

To this end, for a contradiction, assume that F is non-empty. Since it is non-empty it must have a least element, m (the natural numbers are bounded below). $P(1)$ is true, and m cannot be 1. Consider $m - 1$, Since $m > 1$ then $m \in \mathbb{N}$ but as it is smaller than the minimum element $m - 1 \notin F$ and so $P(m - 1)$ is true. Then, by our assumptions $P(m)$ must also be true, meaning that m cannot be in F. This contradicts m being the least element of F, and so we conclude that F must be empty.

> **Remark**
> The analogy of toppling dominoes is often used to describe how mathematical induction works. Once the initial domino has been toppled, each domino is toppled in turn by the preceding domino.

When writing a proof by induction we always write out the statement $P(n)$ we are trying to prove and then do the following four steps:

Step 1: Show that the result is true for a base case (often $n = 1$).
Step 2: Assume the result is true for $n = k$, *i.e.* $P(k)$ is true.
Step 3: Show that $P(k)$ being true implies that $P(k + 1)$ is also true.
Step 4: Write a concluding statement.

The rest of this chapter explores the use of mathematical induction in different situations.

1 Induction for Series

We start by proving the result demonstrated in 1575 by Francesco Maurolico in his "Arithmeticorum libri duo".

Example 6.1
Prove, by induction that,

$$\sum_{r=1}^{n}(2r - 1) = n^2 \tag{6.1}$$

Solution:
We proceed in the usual way.
Let $P(n)$ be the statement "$\sum_{r=1}^{n}(2r - 1) = n^2$."

Step 1: We first check the base case, when $n = 1$.

$$LHS = \sum_{r=1}^{1}(2r - 1)$$
$$= 1,$$
$$RHS = 1^2$$
$$= 1.$$

Hence, $P(n)$ is true for the base case $n = 1$.

Step 2: We assume that $P(n)$ is true for $n = k$, that is, we assume that

$$\sum_{r=1}^{k}(2r-1) = k^2.$$

Step 3: We now seek to show that if $P(k)$ is true then $P(k+1)$ must also be true. It often helps to write out the expression we wish to show. In this case, we wish to show that

$$\sum_{r=1}^{k+1}(2r-1) = (k+1)^2.$$

To this end,

$$\begin{aligned}
\sum_{r=1}^{k+1}(2r-1) &= \sum_{r=1}^{k}(2r-1) + (2(k+1)-1) \\
&= k^2 + (2k+2-1) \\
&= k^2 + 2k + 1 \\
&= (k+1)^2.
\end{aligned}$$

Where on the second line the assumed truth of $P(k)$ has been used. Hence, if the summation formula is true for $n = k$ it is also true for $n = k+1$.

Step 4: Since the summation formula (6.1) is true for $n = 1$, and if true for $n = k$ then also true for $n = k+1$, it is true for all $n \geq 1$ by the principle of mathematical induction.
Hence, $P(n)$ has been proved $\forall n \in \mathbb{N}$.

Proof Tip

We begin Step 3 by splitting the summation up to $n = k+1$ into a summation up to $n = k$ plus an extra term. This is a standard approach when proving summation results by induction. In the second line of Step 3 we have used the inductive hypothesis to write $\sum_{r=1}^{k}(2r-1)$ as k^2.

Example 6.2

Prove, by induction that,

$$\sum_{i=0}^{n} 2^i = 2^{n+1} - 1 \tag{6.2}$$

Solution:
We proceed in the usual way.

Let $P(n)$ be the statement "$\sum_{i=0}^{n} 2^i = 2^{n+1} - 1$."

Step 1: We first check the base case, when $n = 1$.

$$LHS = \sum_{r=0}^{0} 2^i$$
$$= 1,$$
$$RHS = 2^1 - 1$$
$$= 1.$$

Hence, $P(n)$ is true for the base case $n = 1$.

Step 2: We assume that $P(n)$ is true for $n = k$, that is, we assume that

$$\sum_{i=0}^{k} 2^i = 2^{k+1} - 1.$$

Step 3: We now seek to show that if $P(k)$ is true then $P(k + 1)$ must also be true. It often helps to write out the expression we wish to show. In this case, we wish to show that

$$\sum_{i=0}^{k+1} 2^i = 2^{k+2} - 1$$

To this end,

$$\sum_{i=0}^{k+1} 2^i = \sum_{i=0}^{k} 2^i + 2^{k+1}$$
$$= 2^{k+1} - 1 + 2^{k+1}$$
$$= 2 \cdot 2^{k+1} - 1$$
$$= 2^{k+2} - 1.$$

Hence, if the summation formula is true for $n = k$ it is also true for $n = k + 1$.

Step 4: Since the summation formula (6.2) is true for $n = 0$, and if true for $n = k$ then also true for $n = k + 1$, it is true for all $n \geq 0$ by the principle of mathematical induction.

Hence, $P(n)$ has been proved $\forall n \in \mathbb{N}_0$.

We need to take extra care when the upper limit of our summation is not n; it is important in Step 3 that we add in the appropriate number of terms as shown in the example below.

Example 6.3

Prove, by induction, that

$$\sum_{r=1}^{2n}(r^2 + r) = \frac{4}{3}n(n+1)(2n+1).$$

Solution:

Let $P(n)$ be the statement "$\sum_{r=1}^{2n}(r^2 + r) = \frac{4}{3}n(n+1)(2n+1)$".

Step 1: We first check the base case, when $n = 1$.

$$LHS = \sum_{r=1}^{2}(r^2 + r)$$
$$= (1+1) + (4+2)$$
$$= 8$$
$$RHS = \frac{4}{3}(1)(1+1)(2+1)$$
$$= \frac{4}{3} \times 1 \times 2 \times 3$$
$$= 8.$$

Hence, $P(n)$ is true for the base case $n = 1$.

Step 2: We assume that $P(n)$ is true for $n = k$, that is, we assume that

$$\sum_{r=1}^{2k}(r^2 + r) = \frac{4}{3}k(k+1)(2k+1).$$

Step 3: We now seek to show that if $P(k)$ is true then $P(k+1)$ must also be true. It often helps to write out the expression we wish to show. In this case, we wish to show that

$$\sum_{r=1}^{2(k+1)}(r^2 + r) = \frac{4}{3}(k+1)((k+1)+1)(2(k+1)+1)$$

$$= \frac{4}{3}(k+1)(k+2)(2k+3).$$

First note that $\sum_{r=1}^{2(k+1)}(r^2 + r) = \sum_{r=1}^{2(k+1)} r(r+1)$. Then,

$$
\begin{aligned}
\sum_{r=1}^{2(k+1)} (r^2 + r) &= \sum_{r=1}^{2(k+1)} r(r+1) \\
&= \sum_{r=1}^{2k} r(r+1) + (2k+1)(2k+2) + (2k+2)(2k+3) \\
&= \frac{4}{3}k(k+1)(2k+1) + (2k+1)(2k+2) + (2k+2)(2k+3) \\
&= \frac{4}{3}\left[k(k+1)(2k+1) + \frac{3}{4}(2k+1)(2k+2) + \frac{3}{4}(2k+2)(2k+3)\right] \\
&= \frac{4}{3}\left[k(k+1)(2k+1) + \frac{6}{4}(2k+1)(k+1) + \frac{6}{4}(k+1)(2k+3)\right] \\
&= \frac{4}{3}(k+1)\left[k(2k+1) + \frac{6}{4}(2k+1) + \frac{6}{4}(2k+3)\right] \\
&= \frac{4}{3}(k+1)\left[k(2k+1) + 3k + \frac{3}{2} + 3k + \frac{9}{2}\right] \\
&= \frac{4}{3}(k+1)\left[2k^2 + k + 3k + 3k + \frac{3}{2} + \frac{9}{2}\right] \\
&= \frac{4}{3}(k+1)\left[2k^2 + 7k + 6\right] \\
&= \frac{4}{3}(k+1)(k+2)(2k+3)
\end{aligned}
$$

Hence, if the summation formula is true for $n = k$ it is also true for $n = k+1$.

Step 4: Since the summation formula (6.2) is true for $n = 0$, and if true for $n = k$ then also true for $n = k+1$, it is true for all $n \geq 0$ by the principle of mathematical induction.

Hence, $P(n)$ has been proved $\forall n \in \mathbb{N}_0$.

Proof Tip

In Step 3 of Example 6.2 our knowledge of what $P(K+1)$ should be was invaluable. In the 5th line of this step we expressed $2k + 2$ as $2(k+1)$ in two places so there was a common factor of $(k+1)$ which we could factorise out - we knew to do this because the statement for $P(k+1)$ contained a factor of $(k+1)$.

Exercise 6.1

Q1. Prove, by induction,

$$\sum_{r=1}^{n} r = \frac{1}{2}n(n+1);$$

Q2. Prove, by induction,

$$\sum_{r=1}^{n} r^2 = \frac{1}{6}n(n+1)(2n+1);$$

Q3. Prove, by induction,

$$\sum_{r=1}^{n} r^3 = \frac{1}{4}n^2(n+1)^2;$$

Q4. Prove, by induction,

$$\sum_{r=1}^{n} r(r-1) = \frac{1}{3}(n-1)n(n+1);$$

Q5. Prove, by induction,

$$\sum_{r=1}^{n} 3^r = \frac{3}{2}(3^n - 1);$$

Q6. Prove,

$$\sum_{r=1}^{n} r(r!) = (n+1)! - 1;$$

Q7. Prove, by induction,

$$\sum_{r=1}^{n} r^2(r+1) = \frac{1}{12}n(n+1)(n+2)(3n+1);$$

Q8. Prove, by induction,

$$\sum_{r=1}^{n} \frac{2}{(r+1)(r+2)} = \frac{n}{n+2};$$

Q9. Prove, by induction,

$$\sum_{r=1}^{2n} \frac{1}{(r+2)(r+3)} = \frac{2n}{6n+9};$$

Q10. Let $x \in \mathbb{R}$, such that $x \neq -1$ or $x \neq 1$. Prove, by induction on $n \in \mathbb{N}$ that

$$\sum_{r=1}^{n} \frac{x^{2^{r-1}}}{1 - x^{2^r}} = \frac{1}{1-x} - \frac{1}{1 - x^{2^n}}.$$

.2   Induction for Matrices

Mathematical induction provides a very nice way of proving properties of powers of matrices. The technique is very similar to that for series, however we now use the multiplication of matrices instead of the summation of series.

Example 6.4

Prove, by induction that, for all $n \in \mathbb{N}$,

$$\begin{pmatrix} 3 & 0 \\ 0 & 4 \end{pmatrix}^n = \begin{pmatrix} 3^n & 0 \\ 0 & 4^n \end{pmatrix}. \tag{6.3}$$

Solution:

Let $P(n)$ be the statement " $\begin{pmatrix} 3 & 0 \\ 0 & 4 \end{pmatrix}^n = \begin{pmatrix} 3^n & 0 \\ 0 & 4^n \end{pmatrix}$ ".

Step 1: Let $n = 1$, then,

$$LHS = \begin{pmatrix} 3 & 0 \\ 0 & 4 \end{pmatrix}^1$$

$$= \begin{pmatrix} 3 & 0 \\ 0 & 4 \end{pmatrix},$$

$$RHS = \begin{pmatrix} 3^1 & 0 \\ 0 & 4^1 \end{pmatrix}$$

$$= \begin{pmatrix} 3 & 0 \\ 0 & 4 \end{pmatrix}$$

Thus, the left hand side is equal to the right hand side and the basis step for the induction argument has been shown.

Step 2: We assume that $P(k)$ is true, that is,

$$\begin{pmatrix} 3 & 0 \\ 0 & 4 \end{pmatrix}^k = \begin{pmatrix} 3^k & 0 \\ 0 & 4^k \end{pmatrix}$$

Step 3: Assuming that $P(k)$ is true, we now wish to show that the truth of $P(k+1)$ immediately follows. We wish to show that,

$$\begin{pmatrix} 3 & 0 \\ 0 & 4 \end{pmatrix}^{k+1} = \begin{pmatrix} 3^{k+1} & 0 \\ 0 & 4^{k+1} \end{pmatrix}$$

To this end,

$$\begin{pmatrix} 3 & 0 \\ 0 & 4 \end{pmatrix}^{k+1} = \begin{pmatrix} 3 & 0 \\ 0 & 4 \end{pmatrix}^{k} \begin{pmatrix} 3 & 0 \\ 0 & 4 \end{pmatrix}$$

$$= \begin{pmatrix} 3^k & 0 \\ 0 & 4^k \end{pmatrix} \begin{pmatrix} 3 & 0 \\ 0 & 4 \end{pmatrix} \quad \text{by inductive hypothesis}$$

$$= \begin{pmatrix} 3 \cdot 3^k & 0 \\ 0 & 4 \cdot 4^k \end{pmatrix}$$

$$= \begin{pmatrix} 3^{k+1} & 0 \\ 0 & 4^{k+1} \end{pmatrix}$$

Hence $P(k+1)$ is also true.

Step 4: Since the matrix product formula (6.3) is true for $n = 1$, and if true for $n = k$ then also true for $n = k + 1$, it is also true for all $n \in \mathbb{N}$ by the principle of mathematical induction.

Proof Tip
Any induction proof concerning powers of matrices will use $A^{k+1} = A^k A$.

Example 6.5
Prove, by induction that, for all $n \in \mathbb{N}$,

$$\begin{pmatrix} 1 & a \\ 0 & 1 \end{pmatrix}^n = \begin{pmatrix} 1 & an \\ 0 & 1 \end{pmatrix}. \tag{6.4}$$

Solution:
Let $P(n)$ be the statement "$\begin{pmatrix} 1 & a \\ 0 & 1 \end{pmatrix}^n = \begin{pmatrix} 1 & an \\ 0 & 1 \end{pmatrix}$".

Step 1: Let $n = 1$, then,

$$LHS = \begin{pmatrix} 1 & a \\ 0 & 1 \end{pmatrix}^1$$

$$= \begin{pmatrix} 1 & a \\ 0 & 1 \end{pmatrix},$$

$$RHS = \begin{pmatrix} 1 & a \times 1 \\ 0 & 1 \end{pmatrix}$$

$$= \begin{pmatrix} 1 & a \\ 0 & 1 \end{pmatrix}$$

Thus, the left hand side is equal to the right hand side and the basis step for the induction argument has been shown.

Step 2: We assume that $P(k)$ is true, that is,

$$\begin{pmatrix} 1 & a \\ 0 & 1 \end{pmatrix}^k = \begin{pmatrix} 1 & ak \\ 0 & 1 \end{pmatrix}.$$

Step 3: Assuming that $P(k)$ is true, we now wish to show that the truth of $P(k+1)$ immediately follows. We wish to show that,

$$\begin{pmatrix} 1 & a \\ 0 & 1 \end{pmatrix}^{k+1} = \begin{pmatrix} 1 & a(k+1) \\ 0 & 1 \end{pmatrix}.$$

Consider,

$$\begin{pmatrix} 1 & a \\ 0 & 1 \end{pmatrix}^{k+1} = \begin{pmatrix} 1 & a \\ 0 & 1 \end{pmatrix}^k \begin{pmatrix} 1 & a \\ 0 & 1 \end{pmatrix}$$

$$= \begin{pmatrix} 1 & ak \\ 0 & 1 \end{pmatrix} \begin{pmatrix} 1 & a \\ 0 & 1 \end{pmatrix} \qquad \text{by inductive hypothesis}$$

$$= \begin{pmatrix} 1 \times 1 + ak \times 0 & 1 \times a + ak \times 1 \\ 0 \times 1 + 1 \times 0 & 0 \times a + 1 \times 1 \end{pmatrix}$$

$$= \begin{pmatrix} 1 & a + ak \\ 0 & 1 \end{pmatrix}$$

$$= \begin{pmatrix} 1 & a(k+1) \\ 0 & 1 \end{pmatrix}.$$

Hence $P(k+1)$ is also true if $P(k)$ is true.

Step 4: Since the matrix product formula (6.4) is true for $n = 1$, and if true for $n = k$ then also true for $n = k + 1$, it is also true for all $n \in \mathbb{N}$ by the principle of mathematical induction.

Exercise 6.2

Q1. Prove, by induction, that for the matrix $A = \begin{pmatrix} 1 & 0 \\ 1 & 1 \end{pmatrix}$,

$$A^n = \begin{pmatrix} 1 & 0 \\ n & 1 \end{pmatrix}$$

Q2. Prove, by induction, that for the matrix $A = \begin{pmatrix} 1 & 0 & 1 \\ 0 & 1 & 0 \\ 0 & 1 & 1 \end{pmatrix}$,

$$A^n = \frac{1}{2} \begin{pmatrix} 2 & n^2 - n & 2n \\ 0 & 2 & 0 \\ 0 & 2n & 2 \end{pmatrix}$$

Q3. Prove, by induction, that for the matrix $A = \begin{pmatrix} \cos(\theta) & \sin(\theta) \\ -\sin(\theta) & \cos(\theta) \end{pmatrix}$,

$$A^n = \begin{pmatrix} \cos(n\theta) & \sin(n\theta) \\ -\sin(n\theta) & \cos(n\theta) \end{pmatrix}$$

Q4. Prove, by induction, that for the matrix $A = \begin{pmatrix} \cosh^2(\theta) & \cosh^2(\theta) \\ -\sinh^2(\theta) & -\sinh^2(\theta) \end{pmatrix}$,

$$A^n = A$$

Q5. Prove, by induction, that,

$$\begin{pmatrix} 10 & -1 \\ 8 & 1 \end{pmatrix}^n = \frac{1}{7} \begin{pmatrix} -2^n + 8 \times 9^n & 2^n - 9^n \\ -8(2^n - 9^n) & 2^{n+3} - 9^n \end{pmatrix}.$$

Q6. Prove, by induction, that,

$$\begin{pmatrix} 2 & 0 \\ 0 & 2 \end{pmatrix}^n = 2^n \begin{pmatrix} 1 & 0 \\ 0 & 1 \end{pmatrix}.$$

Q7. (a) Prove, by induction, that,

$$\begin{pmatrix} 1 & 1 \\ 1 & 1 \end{pmatrix}^n = \begin{pmatrix} 2^{n-1} & 2^{n-1} \\ 2^{n-1} & 2^{n-1} \end{pmatrix}.$$

(b) Suggest a formula for

$$\begin{pmatrix} 1 & 1 & 1 \\ 1 & 1 & 1 \\ 1 & 1 & 1 \end{pmatrix}^n,$$

and prove it holds.

.3 Induction for Divisibility

It is often important to know if a particular function always generates values that are divisible by a given integer. For example, consider the sequence defined by $f(n) = 4^n - 1$; looking at the first few values we have,

n	$f(n)$
1	3
2	15
3	63
4	255
5	1023
6	4095

It appears that every value of $f(n)$ is divisible by 3 but as we cannot evaluate every possible $f(n)$ we need a proof to enable us to have belief in this conjecture. Proof by induction can also be used here to prove that $f(n)$ is divisible by 3 for all n.

Remark

If a function $f(n)$ is divisible by 3 then we may write $3|f(n)$, which is read "3 divides $f(n)$".

Example 6.6

Prove that $f(n) = 9^n - 1$ is divisible by 8 for all $n \in \mathbb{N}$.

Solution:
Let $P(n)$ be the statement "$8 \mid 9^n - 1$".

Step 1: When $n = 1$, we have

$$9^1 - 1 = 9 - 1$$
$$= 8,$$

which is certainly divisible by 8 so the statement is true for $n = 1$.

Step 2: We assume the $P(k)$ is true, and so $f(k) = 9^k - 1$ is divisible by 8.

Step 3: We wish to show that $P(k) \Rightarrow P(k+1)$. We begin with $9^{k+1} - 1$ and attempt to factor out a $9^k - 1$. To this end,

$$9^{k+1} - 1 = 9(9^k) - 1$$
$$= 9(9^k - 1) + 9 - 1$$
$$= 9(9^k - 1) + 8.$$

By our inductive hypothesis the term $9(9^k - 1)$ in the last line above must be divisible by 8. Also, 8 is clearly divisible by 8, and so $9^{k+1} - 1$ must be divisible by 8 as it is the sum of two terms which are each divisible by 8.

Step 4: We have shown the statement to be true for $n = 1$ and if it is true for $n = k$ then it must also be true for $n = k + 1$. Therefore, by the principle of mathematical induction, the statement must be true for all $n \in \mathbb{N}$.

Proof Tip

We could have proceeded slightly differently with Step 2 and Step 3 as shown below.

Step 2: We assume the $P(k)$ is true, and so we may write,

$$9^k - 1 = 8A,$$

for some integer A.

Step 3

$$9^{k+1} - 1 = 9(9^k) - 1$$
$$= 9(9^k - 1) + 9 - 1$$
$$= 9(8A) + 8,$$
$$= 8(9A + 1).$$

From the last line above it is clear that $4^{k+1} - 1$ is divisible by 3.

This notational approach has its advantages, and is particularly helpful to students when they first encounter divisibility proofs.

Proof Tip

It is sometimes suggested that we should do the following:

Step 3: We consider $f(k + 1) - f(k)$,

$$f(k + 1) - f(k) = 9^{k+1} - 1 - (9^k - 1)$$
$$= 9 \cdot 9^k - 1 - 9^k + 1$$
$$= 8 \cdot 9^k.$$

We now know that the difference between $f(k+1)$ and $f(k)$ is a multiple of 8. Hence, as we assumed $f(k)$ was a multiple of 8, $f(k+1)$ must also be.

The authors do not recommend this as an appropriate method when tackling divisibility proofs for the following reasons:

- The expression $f(k+1) - f(k)$ is often not as nice to manipulate as in the above example, and sometimes includes a multiple of $f(k)$ which is rather ugly.
- Students often seem to struggle with this approach and find the method shown in Example 6.4 easier to understand and then replicate.
- This approach feels at odds with other forms of induction where we seek to start with an expression for $f(k+1)$ and then rearrange to express it in terms of $f(k)$, whence we can apply the inductive hypothesis.

Remark

It should be noted that this result could be shown in other ways, for example, consider the alternative proofs.

Alternative 1:

$$9^n - 1 = 3^{2n} - 1$$
$$= (3^n - 1)(3^n + 1).$$

This means that $9^n - 1$ is the product of two consecutive even numbers (since 3^n is odd). Hence one is divisible by 2 and the other by 4 and $9^n - 1$ is divisible by 8.

Alternative 2: Using the binomial theorem,

$$9^n = (8 + 1)^n$$
$$= 8^n + \binom{n}{1}8^{n-1}1^1 + \binom{n}{2}8^{n-2}1^2 + \cdots + \binom{n}{n-1}8^1 1^{n-1} + 1^n$$
$$= 8\left(8^{n-1} + \binom{n}{1}8^{n-2}1^1 + \binom{n}{2}8^{n-3}1^2 + \cdots + \binom{n}{n-1}\right) + 1$$
$$= 8m + 1 \qquad \text{for } m \text{ a positive integer.}$$
$$\Rightarrow \quad 9^n - 1 = 8m.$$

We conclude that $9^n - 1$ is a multiple of 8 for any integer n.

Example 6.7

Prove that $f(n) = 9^n - 2^n$ is divisible by 7 for all $n \in \mathbb{N}$.

Solution:

Let $P(n)$ be the statement "$7 \mid 9^n - 2^n$.

Step 1: When $n = 1$, we have

$$9^1 - 2^1 = 9 - 2$$
$$= 7$$

which is certainly divisible by 7 so the statement is true for $n = 1$.

Step 2: We assume the $P(k)$ is true, and so $f(k) = 9^k - 2^k$ is divisible by 7.

Step 3: We wish to show that $P(k) \Rightarrow P(k+1)$. We begin with $9^{k+1} - 2^{k+1}$ and attempt to factor out a $9^k - 2^k$. To this end,

$$9^{k+1} - 2^{k+1} = 9 \cdot 9^k - 2 \cdot 2^k$$
$$= 9 \cdot 9^k - 2 \cdot 2^k + 9 \cdot 2^k - 9 \cdot 2^k$$
$$= 9(9^k - 2^k) - 2 \cdot 2^k + 9 \cdot 2^k$$
$$= 9(9^k - 2^k) + 7 \cdot 2^k$$

By our inductive hypothesis the term $9(9^k - 2^k)$ in the last line above must be divisible by 7. Also, $7 \cdot 2^k$ is divisible by 7, and so $9^{k+1} - 2^{k+1}$ must be divisible by 7 as it is the sum of two terms which are each divisible by 7.

Step 4: We have shown the statement to be true for $n = 1$ and if it is true for $n = k$ then it must also be true for $n = k + 1$. Therefore, by the principle of mathematical induction, the statement must be true for all $n \in \mathbb{N}$.

Proof Tip

In the second line of Step 3 we used the classic "proof trick" of adding zero to an expression in a way that is useful. In this case we added $0 = 9 \cdot 2^k - 9 \cdot 2^k$ so that the $-9 \cdot 2^k$ could be combined with the $9 \cdot 9^k$ to enable a factorisation of $f(k)$.

Remark

If, as is sometimes suggested, we had considered $f(k+1) - f(k)$ in the inductive step (Step 3) we would have, in this case,

$$f(k+1) - f(k) = 9^{k+1} - 2^{k+1} - (9^k - 2^k)$$
$$= 9 \cdot 9^k - 2 \cdot 2^k - 9^k + 2^k$$
$$= 8 \cdot 9^k - 2^k$$
$$= 8(9^k - 2^k) + 7 \cdot 2^k$$
$$= 8f(k) + 7 \cdot 2^k.$$

Example 6.8

Prove that $f(n) = n^3 + 5n + 6$ is divisible by 3 for all $n \in \mathbb{N}$.

Solution:

Let $P(n)$ be the statement "$3 \mid n^3 + 5n + 6$.

Step 1: When $n = 1$, we have

$$1^3 + 5 \times 1 + 6 = 12$$
$$= 3 \times 4$$

which is certainly divisible by 3 so the statement is true for $n = 1$.

Step 2: We assume the $P(k)$ is true, and so $f(k) = k^3 + 5k + 6$ is divisible by 3.

Step 3: We wish to show that $P(k) \Rightarrow P(k+1)$. We begin with $(k+1)^3 + 5(k+1) + 6$ and attempt to factor out a $k^3 + 5k + 6$. To this end,

$$(k+1)^3 + 5(k+1) + 6 = k^3 + 3k^2 + 3k + 1 + 5k + 5 + 6$$
$$= (k^3 + 5k + 6) + 3k^2 + 3k + 6$$
$$= (k^3 + 5k + 6) + 3(k^2 + k + 2).$$

By our inductive hypothesis the term $(k^3 + 5k + 6)$ in the last line above must be divisible by 3. Also, $3(k^2 + k + 2)$ is divisible by 3, and so $(k+1)^3 + 5(k+1) + 6$ must be divisible by 3 as it is the sum of two terms which are each divisible by 3.

Step 4: We have shown the statement to be true for $n = 1$ and if it is true for $n = k$ then it must also be true for $n = k + 1$. Therefore, by the principle of mathematical induction, the statement must be true for all $n \subset \mathbb{N}$.

Example 6.9

Prove, by induction, that $x^n - 1$ is divisible by $(x - 1)$.

Solution:

Let $P(n)$ be the statement "$(x - 1) \mid (x^n - 1)$".

Step 1: Consider the case $n = 1$. The polynomial $x^1 - 1 = x - 1$ is certainly divisible by $(x - 1)$, hence $P(1)$ is true.

Step 2: We assume that $P(k)$ is true. This means that we may write,

$$x^k - 1 = (x - 1)p(x),$$

for some polynomial $p(x)$.

Step 3: We wish to show that $(x - 1)$ divides $x^{k+1} - 1$. This amounts to showing that,

$$x^{k+1} - 1 = (x - 1)q(x),$$

for some polynomial $q(x)$.
To this end,

$$
\begin{aligned}
x^{k+1} - 1 &= x(x^k) - 1 \\
&= x(x^k) - x + x - 1 \\
&= x(x^k - 1) + (x - 1) \\
&= x(x - 1)p(x) + (x - 1) \quad \text{(by inductive hypothesis)} \\
&= (x - 1)(xp(x) + 1).
\end{aligned}
$$

Hence $x^{k+1} - 1 = (x - 1)q(x)$ where $q(x) = xp(x) + 1$ and we have shown that if $P(k)$ is true then so is $P(k + 1)$.

Step 4: We have shown the statement to be true for $n = 1$ and if it is true for $n = k$ then it must also be true for $n = k + 1$. Therefore, by the principle of mathematical induction, the statement must be true for all $n \in \mathbb{N}$.

Proof Tip

In line two of the manipulations in Step 3 we have again seen the "trick" of adding 0 (in this case $-x + x$) to allow us to proceed with the proof.

Exercise 6.3

Q1. Prove that $4^n - 1$ is divisible by 3 for all $n \in \mathbb{N}$.
Q2. Prove that $7^n - 7$ is divisible by 6 for every integer $n \geq 1$.
Q3. Prove that $5^n + 5$ is divisible by 3 for every even integer $n \geq 2$.
Q4. Prove that $23^n - 1$ is a multiple of 11 for every integer $n \geq 1$.

Q5. Prove that $8^n - 3^n$ is divisible by 5 or every integer $n \geq 1$.

Q6. Prove that $3^{3n} + 2^{n+2}$ is divisible by 5 for every integer $n \geq 0$.

Q7. Prove that $2^{6n} + 3^{2n-2}$ is divisible by 5 for every integer $n \geq 1$.

Q8. Prove that $5^{2n+1} + 6^{3n+2} + 7$ is divisible by 12 for every integer $n \geq 0$.

Q9. Prove that $x + 1$ is a factor of $x^n + 1$ for all integer $n \geq 1$.

Q10. Prove that $2^{2n} - 1$ is divisible by 3 for all integers $n \geq 1$.

Q11. Prove that $2^{n+2} + 3^{2n+1}$ is divisible by 7 for all integers $n \geq 1$.

Q12. Prove that $5^n + 3$ is divisible by 4 for all integers $n \geq 1$.

.4 Induction for Inequalities

Example 6.10

Prove by induction that

$$2^n > n \qquad \forall n \geq 1 \tag{6.5}$$

Solution:

Let $P(n)$ be the statement "$2^n > n \quad \forall n \geq 1$".

Step 1: We check the base case, here this is $n = 1$.

$$LHS = 2^1$$
$$= 2,$$
$$RHS = 1.$$

Hence, $P(n)$ is true for the base case $n = 1$.

Step 2: We assume that $P(n)$ is true for $n = k$, that is, we assume that

$$2^k > k.$$

Step 3: We now seek to show that if $P(k)$ is true then $P(k+1)$ must also be true. It often helps to write out the expression we wish to show. In this case, we wish to show that

$$2^{k+1} > k + 1.$$

To this end,

$$2^{k+1} = 2(2^k) \qquad \text{(by inductive hypothesis)}$$
$$> 2k$$
$$> k + 1,$$

In the last line we have used $k > 1$ implies that $2k > k + 1$ to get the result.

Step 4: Since inequality (6.5) is true for $n = 1$, and if true for $n = k$ then also true for $n = k + 1$, it is also true for all $n \geq 1$ by the principle of mathematical induction.

Hence, $P(n)$ has been proven for all $n > 1$.

Frequently we are interested in inequalities involving factorials.

Example 6.11

Given that n is an integer, prove by induction that

$$n! > 2^n \qquad \forall n \geq 4. \tag{6.6}$$

Solution:

Let $P(n)$ be the statement "$n! > 2^n \quad \forall n \geq 4$".

Step 1: We check the base case, here this is $n = 4$.

$$\begin{aligned}
LHS &= 4! \\
&= 24, \\
RHS &= 2^4 \\
&= 16
\end{aligned}$$

Hence, $P(n)$ is true for the base case $n = 4$.

Step 2: We assume that $P(n)$ is true for $n = k$, that is, we assume that

$$k! > 2^k.$$

Step 3: We now seek to show that if $P(k)$ is true then $P(k + 1)$ must also be true. It often helps to write out the expression we wish to show. In this case, we wish to show that

$$(k + 1)! > 2^{k+1}.$$

To this end, we begin with our result for $P(k)$

$$\begin{aligned}
k! &> 2^k \\
\Rightarrow \quad (k + 1)k! &> (k + 1)2^k \\
&> 2 \times 2^k \\
&= 2^{k+1}
\end{aligned}$$

In the third line we have used $k > 4$ implies that $k + 1 > 5$ and so certainly larger than 2, to get the result.

Step 4: Since inequality (6.6) is true for $n = 4$, and if true for $n = k$ then also true for $n = k + 1$, it is also true for all $n \geq 4$ by the principle of mathematical induction.

Hence, $P(n)$ has been proven for all $n \geq 4$.

Proof Tip

Note that in the above example we started with $P(k)$ and manipulated both sides of the inequality so that the left hand side became the left hand side of $P(k+1)$.

Example 6.12

Let $n \in \mathbb{N}$. Prove that for all n,

$$n^2 + 5n + 3 < 15n^2. \tag{6.7}$$

Solution:

Let $P(n)$ be the statement "$n^2 + 5n + 3 < 15n^2 \quad \forall n \geq 1$".

Step 1: We check the base case, here this is $n = 1$.

$$
\begin{aligned}
LHS &= 1^2 + 5 \times 1 + 3 \\
&= 9, \\
RHS &= 15 \times 1^2 \\
&= 15.
\end{aligned}
$$

Hence, $P(n)$ is true for the base case $n = 1$.

Step 2: We assume that $P(n)$ is true for $n = k$, that is, we assume that

$$k^2 + 5k + 3 < 15k^2.$$

Step 3: We now seek to show that if $P(k)$ is true then $P(k+1)$ must also be true. It often helps to write out the expression we wish to show. In this case, we wish to show that

$$(k+1)^2 + 5(k+1) + 3 < 15(k+1)^2.$$

To this end,

$$
\begin{aligned}
(k+1)^2 + 5(k+1) + 3 &= k^2 + 2k + 1 + 5k + 5 + 3 \\
&= (k^2 + 5k + 3) + 2k + 6 \\
&< 15k^2 + 2k + 6 \quad \text{(by inductive hypothesis)} \\
&< 15k^2 + 30k + 6 \quad \text{Since } 2k < 30k \\
&< 15k^2 + 30k + 15 \quad \text{Since } 4 < 15 \\
&= 15(k^2 + 2k + 1) \\
&= 15(k+1)^2.
\end{aligned}
$$

Hence if $P(k)$ is true then so is $P(k+1)$.

Step 4: Since inequality (6.7) is true for $n = 1$, and if true for $n = k$ then also true for $n = k+1$, it is also true for all $n \geq 1$ by the principle of mathematical induction.

Proof Tip
The authors found that in Step 3, after applying the inductive hypothesis it was not clear how to proceed. At this point they considered what the right hand side of $P(k+1)$ would be and used this to guide them. This is often helpful, however, when writing up the proof it should flow logically down the page.

Exercise 6.4
Q1. Prove that $n! > 15n$ for all integers $n \geq 5$.
Q2. Prove that $3^n > 3n$ for all integers $n > 1$.
Q3. Prove, by induction, that $5^n + 9 < 6^n$ for $n \geq 2$.
Q4. Prove, by induction, that $n^2 \geq 2n + 3$ for all $n \geq 3$.
Q5. Prove, by induction, that $n^2 + 8n + 6 < 20n^2$ for all natural numbers n.
Q6. Let $x \in \mathbb{R}$, prove, by induction, that $(1 + x)^n \geq 1 + nx$ for $n \in \mathbb{N}$.
Q7. For $n \in \mathbb{Z}$, $n > 1$ prove that $12^n > 7^n + 5^n$.
Q8. For $n \geq 2$ prove that $2n < (n + 1)!$.
Q9. Prove that $n^3 < 2^n$ for $n \geq 10$.
Q10. Prove for $n \geq 4$, $2n + 2 < 2^n$.

6.5   Induction for Recurrence Relations

Example 6.13
Prove, by induction, that for the sequence defined by the recurrence formula, with $n \in \mathbb{Z}^+$,

$$f(n + 1) = 2f(n) + 3, n \geq 1, \quad f(1) = 1,$$

the closed form solution is

$$f(n) = 2^{n+1} - 3 \tag{6.8}$$

Solution:
Let $P(n)$ be the statement "$f(n) = 2^{n+1} - 3$".

Step 1: From $f(n) = 2^{n+1} - 3$ we obtain $f(1) = 2^2 - 3 = 1$ which is our given first term. Hence $P(1)$ is true.

Step 2: We assume $P(k)$ holds, that is, we assume that $f(k) = 2^{k+1} - 3$.

Step 3: For $P(k + 1)$ we wish to show that $f(k + 1) = 2^{k+2} - 3$; to this end we consider the recurrence formula and apply our inductive hypothesis:

$$\begin{aligned}
f(k + 1) &= 2f(k) + 3 \\
&= 2(2^{k+1} - 3) + 3 \quad \text{(by inductive hypothesis)} \\
&= 2^{k+2} - 6 + 3 \\
&= 2^{k+2} - 3.
\end{aligned}$$

Hence, if $P(k)$ is true then $P(k+1)$ is also true.

Step 4: Since the result is true for $n = 1$, and if true for $n = k$ it is also true for $n = k+1$, it is true for all $n \in \mathbb{Z}^+$ by the principle of mathematical induction.

Example 6.14

A sequence is described by the recurrence formula

$$u_{n+2} = 5u_{n+1} - 6u_n, \quad n \geq 1, \quad u_1 = 1, u_2 = 5.$$

Show that the general formula $u_n = 3^n - 2^n$ is satisfied.

Solution:

Let $P(n)$ be the statement "$u_n = 3^n - 2^n$".

Step 1: We have to check that the formula for u_n gives the correct values for u_1 and u_2.

When $n = 1$, $u_1 = 3^1 - 2^1 = 1$.

When $n = 2$, $u_2 = 3^2 - 2^2 = 5$.

Hence $P(1)$ and $P(2)$ are true.

Step 2: We now assume $P(k)$ and $P(k+1)$ holds. That is, we assume that $u_k = 3^k - 2^k$ and $u_{k+1} = 3^{k+1} - 2^{k+1}$.

Step 3: For $P(k+2)$ we wish to show that $u_{k+2} = 3^{k+2} - 2^{k+2}$; to this end we consider the recurrence formula and apply our inductive hypotheses:

$$
\begin{aligned}
u_{k+2} &= 5u_{k+1} - 6u_k \\
&= 5(3^{k+1} - 2^{k+1}) - 6(3^k - 2^k) \\
&= 5 \times 3^{k+1} - 5 \times 2^{k+1} - 2 \times 3^{k+1} + 3 \times 2^{k+1} \\
&= 3 \times 3^{k+1} - 2 \times 2^{k+1} \\
&= 3^{k+2} - 2^{k+2},
\end{aligned}
$$

as required. Hence, if $P(k)$ and $P(k+1)$ are true then $P(k+2)$ is also true.

Step 4: Since the result is true for $n = 1$, and if true for $n = k$ and $n = k+1$ it is also true for $n = k+2$, it is true for all $n \geq 1$ by the principle of mathematical induction.

Exercise 6.5

Q1. A sequence is described by the recurrence formula

$$u_{n+1} = 2u_n + 4, \quad n \geq 1, \quad u_1 = 1.$$

Show that the general formula $u_n = 5 \times 2^{n-1} - 4$ is satisfied.

Q2. A sequence is described by the recurrence formula

$$u_{n+1} = 3u_n - 3, \quad n \geq 1, \quad u_1 = 2.$$

Show that the general formula $u_n = \frac{1}{6}(3^n + 9)$ is satisfied.

Q3. A sequence is described by the recurrence formula

$$u_{n+1} = 4u_n + 6, \quad n \geq 1, \quad u_1 = 2.$$

Show that the general formula $u_n = 4^n - 2$ is satisfied.

Q4. A sequence is described by the recurrence formula

$$u_{n+1} = 3u_n + 1, \quad n \geq 1, \quad u_1 = 5.$$

Show that the general formula $u_n = \frac{1}{6}(11 \times 3^n - 3)$ is satisfied.

6.6   Applications of Proof by Induction

There are many applications of induction that are not restricted to the more formulaic A-Level style questions that have been presented in the previous sections.

Indeed, some important results from other areas of school and college mathematics syllabi can be proved by induction. In the First Year content the binomial theorem for integer n is studied.

Theorem 6.2 — Binomial Theorem for Positive Integer n

Let $a, b \in \mathbb{R}$, and $n \in \mathbb{N}$. Then the expansion of $(a + bx)^n$ is given by

$$(a + bx)^n = \binom{n}{0}a^n + \binom{n}{1}a^{n-1}bx + \binom{n}{2}a^{n-2}b^2x^2 + \ldots + \binom{n}{n-1}ab^{n-1}x^{n-1}$$

$$+ \binom{n}{n}b^nx^n,$$

$$= \sum_{r=0}^{n} \binom{n}{r}a^{n-r}b^rx^r.$$

Now we are familiar with proof by induction we can provide a rigorous proof for this theorem, instead of the heuristic explanation based on spotting the pattern of the first few expansions that is often presented during an A-Level course.

Letting $P(n)$ be the statement

$$\text{``}(a + bx)^n = \sum_{r=0}^{n} \binom{n}{r}a^{n-r}b^rx^n\text{'',}$$

we proceed with the usual four steps in the following way.

Step 1: For the basis of our mathematical induction we consider $n = 1$.

$$LHS = (a + bx)^1$$
$$= a + bx,$$

$$RHS = \sum_{r=0}^{1} \binom{1}{r} a^{n-r} b^r x^r$$

$$= \binom{1}{0} a^{1-0} b^0 x^0 + \binom{1}{1} a^{1-1} b^1 x^1$$

$$= a + bx.$$

Hence $P(n)$ is true for $n = 1$.

Step 2: We assume that $P(k)$ is true, that is,

$$(a + bx)^k = \sum_{r=0}^{k} \binom{k}{r} a^{k-r} b^r x^r$$

Step 3: We wish to show that

$$(a + bx)^{k+1} = \sum_{r=0}^{k+1} \binom{k+1}{r} a^{k+1-r} b^r x^r \tag{6.9}$$

With this is mind we consider $P(k+1)$.

$$(a + bx)^{k+1} = (a + bx)^k (a + bx)$$

$$= \left(\sum_{r=0}^{k} \binom{k}{r} a^{k-r} b^r x^r \right) (a + bx) \quad \text{(by inductive hypothesis)}$$

$$= \sum_{r=0}^{k} \binom{k}{r} a^{k-r+1} b^r x^r + \sum_{r=0}^{k} \binom{k}{r} a^{k-r} b^{r+1} x^{r+1}$$

$$= \sum_{r=0}^{k} \binom{k}{r} a^{k+1-r} b^r x^r + \sum_{r=0}^{k} \binom{k}{r} a^{k-r} b^{r+1} x^{r+1}.$$

Comparing the last line above to (6.9) we see that the first summation contains terms of the form we wish to see. To make the second summation fit this form we define $s = r + 1$, which implies that $r = s - 1$. With this we may write the second summation in the following way:

$$\sum_{r=0}^{k} \binom{k}{r} a^{k-r} b^{r+1} x^{r+1} = \sum_{s=1}^{k+1} \binom{k}{s-1} a^{k-s+1} b^s x^s.$$

However, s is just a dummy variable and so we may in fact write (since r is also a dummy variable),

$$\sum_{r=0}^{k} \binom{k}{r} a^{k-r} b^{r+1} x^{r+1} = \sum_{r=1}^{k+1} \binom{k}{r-1} a^{k-r+1} b^r x^r.$$

With this we have,

$$(a+bx)^{k+1} = \sum_{r=0}^{k} \binom{k}{r} a^{k+1-r} b^r x^r + \sum_{r=1}^{k+1} \binom{k}{r-1} a^{k-r+1} b^r x^r$$

$$= \binom{k}{0} a^{k+1} b^0 x^0 + \sum_{r=1}^{k} \binom{k}{r} a^{k+1-r} b^r x^r + \sum_{r=1}^{k} \binom{k}{r-1} a^{k-r+1} b^r x^r$$

$$+ \binom{k}{(k+1)-1} a^{k-(k+1)+1} b^{k+1} x^{k+1},$$

where in the second line we have taken out the $r = 0$ case from the first sum and the $r = k+1$ case from the second sum. Using laws of indices, the fact that $\binom{k}{0} = 1$ and the fact that $\binom{k}{(k+1)-1} = \binom{k}{k} = 1$,

$$(a+bx)^{k+1} = a^{k+1} + \sum_{r=1}^{k} \binom{k}{r} a^{k+1-r} b^r x^r + \sum_{r=1}^{k} \binom{k}{r-1} a^{k-r+1} b^r x^r + b^{k+1} x^{k+1}$$

$$= a^{k+1} + b^{k+1} x^{k+1} \sum_{r=1}^{k} \left[\binom{k}{r} + \binom{k}{r-1} \right] a^{k+1-r} b^r x^r$$

$$= a^{k+1} + b^{k+1} x^{k+1} \sum_{r=1}^{k} \binom{k+1}{r} a^{k+1-r} b^r x^r$$

where we have used Pascal's Identity (see Example 4.18) in the last line of the above. Finally, noting the following,

$$a^{k+1} = \binom{k+1}{k+1} a^{k+1} b^0 x^0,$$

$$b^{k+1} x^{k+1} = \binom{k+1}{0} a^0 b^{k+1} x^{k+1},$$

we have

$$(a+bx)^{k+1} = \sum_{r=0}^{k} \binom{k+1}{r} a^{k+1-r} b^r x^r.$$

Hence, we have the desired form of $P(k+1)$ and have shown that $P(k+1)$ follows from $P(k)$.

Step 4: Since $P(n)$ is true for $n = 1$, and if true for $n = k$ then also true for $n = k + 1$, it is also true for all $n \geq 1$ by the principle of mathematical induction.

Proof Tip

In the above we have proved directly the binomial theorem as often stated. We note that in general the algebra would have been less messy had we proved the statement for $(a+y)^n$ and then let $y = bx$ to obtain the result.

We now present a few other applications of induction, before a few exercises for the reader.

Example 6.15

Prove De Moivre's theorem for a complex number z and integer exponent n:

$$z^n = [r(\cos(\theta) + i\sin(\theta))]^n = r^n(\cos(n\theta) + i\sin(n\theta)) \tag{6.10}$$

Solution:

Without loss of generality we prove De Moivre's Theorem in the case that r, the modulus of z, is 1.

Let $P(n)$ be the statement "$(\cos(\theta) + i\sin(\theta))^n = \cos(n\theta) + i\sin(n\theta)$"

Step 1: For the basis of our mathematical induction we consider $n = 1$.

$$\begin{aligned}
LHS &= (\cos(\theta) + i\sin(\theta))^1 \\
&= \cos(\theta) + i\sin(\theta), \\
RHS &= \cos(1 \times \theta) + i\sin(1 \times \theta) \\
&= \cos(\theta) + i\sin(\theta).
\end{aligned}$$

Hence $P(n)$ is true for $n = 1$.

Step 2:

We assume that $P(k)$ is true, that is,

$$(\cos(\theta) + i\sin(\theta))^k = \cos(k\theta) + i\sin(k\theta)$$

Step 3: We wish to show that

$$(\cos(\theta) + i\sin(\theta))^{k+1} = \cos((k+1)\theta) + i\sin((k+1)\theta$$

With this is mind we consider $P(k+1)$.

$$\begin{aligned}
(\cos(\theta) + i\sin(\theta))^{k+1} &= (\cos(\theta) + i\sin(\theta))^k(\cos(\theta) + i\sin(\theta)) \\
&= (\cos(k\theta) + i\sin(k\theta))(\cos(\theta) + i\sin(\theta)) \\
&= \cos(k\theta)\cos(\theta) + i\cos(k\theta)\sin(\theta) + i\sin(k\theta)\cos(\theta) \\
&\quad - \sin(k\theta)\sin(\theta) \\
&= \cos((k+1)\theta) + i\sin((k+1)\theta)
\end{aligned}$$

where we have used our inductive hypothesis in line two, and Formulae 9.3 in the last line.

Step 4: Since $P(n)$ is true for $n = 1$, and if true for $n = k$ then also true for $n = k+1$, it is also true for all $n \geq 1$ by the principle of mathematical induction.

Hence, $P(n)$ has been proven for all $n \in \mathbf{N}$.

However, the question asked for a proof for all integer n.

When $n = 0$ the result is trivial, since,

$$\begin{aligned}
LHS &= (\cos(\theta) + i\sin(\theta))^0 \\
&= 1, \\
RHS &= \cos(0) + i\sin(0) \\
&= 1
\end{aligned}$$

For negative integers we consider,

$$\begin{aligned}
(\cos(\theta) + i\sin(\theta))^{-n} &= ((\cos(\theta) + i\sin(\theta))^n)^{-1} \\
&= (\cos(n\theta) + i\sin(n\theta))^{-1} \\
&= \cos(-n\theta) + i\sin(-n\theta)
\end{aligned}$$

Hence, De Moivre's Theorem has been shown to be true for all integer exponents n.

Example 6.16

Prove that $\frac{d}{dx}x^n = nx^{n-1}$, for $n \in \mathbb{N}$.

Solution:

Let $P(n)$ be the statement "$\frac{d}{dx}x^n = nx^{n-1}$".

Step 1: When $n = 1$ we have,

$$\begin{aligned}
LHS &= \frac{dx^1}{dx} \\
&= 1, \\
RHS &= 1 \times x^0 \\
&= 1.
\end{aligned}$$

Hence $P(n)$ is true for $n = 1$.

Step 2: We assume that $P(k)$ is true,

$$\frac{d}{dx}x^k = kx^{k-1}$$

Step 3: For $P(k+1)$ we write $x^{k+1} = x^k \cdot x$ and then differentiate using the product rule. We wish to find that

$$\frac{d}{dx}x^{k+1} = (k+1)x^k$$

Proceeding in the way outlined above,

$$\frac{d}{dx}x^{k+1} = \frac{d}{dx}(x \cdot x^k)$$

$$= x \cdot \frac{d}{dx}x^k + \frac{d}{dx}x \cdot x^k$$

$$= x(kx^{k-1}) + x^k \qquad \text{by the inductive hypothesis}$$

$$= kx^k + x^k$$

$$= (k+1)x^k.$$

Step 4: Since $P(n)$ is true for $n = 1$, and if true for $n = k$ then also true for $n = k+1$, it is also true for all $n \geq 1$ by the principle of mathematical induction.
Hence, $P(n)$ has been proven for all $n \in \mathbb{N}$.

We can use proof by induction to find a general formula for the finite sum of the first n terms of a geometric series as shown in the example below.

Example 6.17
Prove,

$$\sum_{r=0}^{n-1} ap^r = \frac{a(1-p^r)}{1-p}$$

Solution:
For this proof we use mathematical induction.
Step 1: We consider $n = 1$,

$$LHS = \sum_{r=0}^{0} ap^r$$

$$= a,$$

$$RHS = \frac{a(1-p^1)}{1-p}$$

$$= a.$$

Hence the result is true for $n = 1$.

Step 2: We assume that $P(k)$ is true,

$$\sum_{r=0}^{k-1} ap^r = \frac{a(1-p^k)}{1-p}.$$

Step 3: We now proceed to show that $P(k) \Rightarrow P(k+1)$, that is we show that

$$\sum_{r=0}^{k} ap^r = \frac{a(1-p^{k+1})}{1-p}.$$

To this end, for $n = k + 1$,

$$\sum_{r=0}^{k} ap^r = \sum_{r=0}^{k-1} ap^r + ap^k$$

$$= \frac{a(1 - p^k)}{1 - p} + ap^k \quad \text{(by inductive hypothesis)}$$

$$= a \left[\frac{(1 - p^k) + p^k(1 - p)}{(1 - p)} \right]$$

$$= a \left[\frac{1 - p^k + p^k - p^{k+1}}{(1 - p)} \right]$$

$$= \frac{a(1 - p^{k+1})}{1 - p}.$$

Hence, if $P(k)$ is true then so is $P(k + 1)$.

Step 4: Since $P(n)$ is true for $n = 1$, and if true for $n = k$ then also true for $n = k + 1$, it is also true for all $n \geq 1$ by the principle of mathematical induction.

Exercise 6.6

Q1. For the function $f(x) = \frac{1}{x}$,
    (a) Find $f'(x)$, $f''(x)$, $f^{(3)}(x)$ and $f^{(4)}(x)$.
    (b) Make a conjecture concerning the nth derivative of $f(x)$.
    (c) Use induction to prove your conjecture.

Q2. Prove, by induction,

$$(\cosh(\theta) + \sinh(\theta))^n \equiv \cosh(n\theta) + \sinh(n\theta)$$

Q3. Prove,

$$\int x^n \, dx = \frac{1}{n+1} x^{n+1} + C$$

for all positive integral values of n.

Q4. Prove,

$$\frac{d^n}{d\theta^n}(\cos(a\theta)) = a^n \cos\left(a\theta + \frac{n\pi}{2}\right)$$

Q5. Define the Fibonacci numbers F_n by the following recursive rule,

$$F_n = F_{n-1} + F_{n-2}, \qquad \text{for } n \geq 2,$$

with initial values $F_0 = 0$ and $F_1 = 1$.
    (a) Find the first 8 terms of the sequence.

(b) Prove the following closed expression for a Fibonacci number, for every integer $n \geq 0$.

$$F_n = \frac{\alpha^n - \beta^n}{\sqrt{5}},$$

where,

$$\alpha = \frac{1 + \sqrt{5}}{2},$$
$$\beta = \frac{1 - \sqrt{5}}{2}.$$

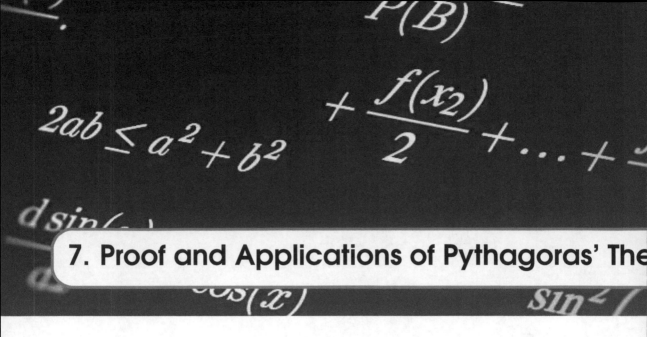

7. Proof and Applications of Pythagoras' The

Pythagoras' Theorem is perhaps the most widely known and recognisable theorem throughout mathematics. It is attributed to Pythagoras of Samos, a Greek philosopher who lived *ca* 570BC-495BC. Although the theorem takes his name, it is unlikely that he was the first to prove it and certainly the result was well know before his time. Figure 7.1 shows an ancient Chinese geometric proof of the theorem, which dates to around the same time as Pythagoras.

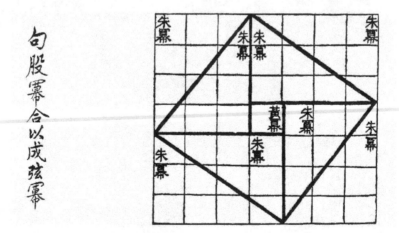

Figure 7.1: Ancient Chinese Geometric Proof of Pythagoras' Theorem.

In this chapter we prove Pythagoras' Theorem in a number of ways, then use the theorem to prove some more complex results met in the A-Level syllabus.

7.1 Proving Pythagoras' Theorem

There are many ways of proving Pythagoras' Theorem. In this section we show three common methods, including the one believed to be used by Pythagoras and one which can be found in Euclid's Elements.

.1 Proof by Pythagoras

The proof shown below is alleged to be the one Pythagoras used to prove the theorem. Figure 9.1 shows a square of side-length c inscribed in a larger square of side-length $a + b$.

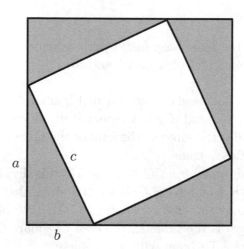

Figure 7.2: A square of side length c inscribed in a larger square.

Using symmetry, the four shaded triangles surrounding the inner square are congruent. Each of the congruent triangles has area $\frac{1}{2}ab$. The smaller and larger squares have area c^2 and $(a + b)^2$, respectively. An alternative expression for the area of the smaller square is the area of the larger square minus the area of the four congruent triangles. Hence

$$
\begin{aligned}
c^2 &= (a + b)^2 - 4\left(\frac{1}{2}ab\right) \\
&= a^2 + b^2 + 2ab - 2ab \\
&= a^2 + b^2.
\end{aligned}
$$

Hence, Pythagoras' theorem is proved.

.2 Proof by Euclid

Euclid's proof relies on a number of results that we are already familiar with from GCSE, or before. They are

(P1) Two triangles ABC and DEF are congruent if $|AB| = |DE|$, $|BC| = |EF|$ and $\angle ABC = \angle DEF$.

(P2) Congruent triangles have the same area.

(P3) Consider a rectangle with vertices $ABCD$ and a triangle with base of length $b = |AB|$ and perpendicular height of length $h = |BC|$. The triangle has the area $\frac{bh}{2}$, half that of the area of rectangle $ABCD$. For clarity, this result is shown in Figure 7.3.

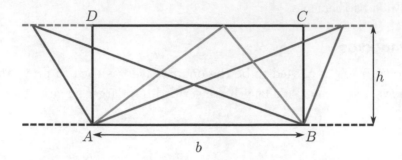

Figure 7.3: Triangles with the same base b and perpendicular height h have half the area of the rectangle with adjacent sides of length b and h.

Consider the right angled triangle ABC shown in Figure 7.4(a) appended with squares with sides of length $|AB|$, $|BC|$ and $|CA|$. Geometrically, Pythagoras' theorem says that the area of the larger square is the same as the sum of the areas of the two smaller squares. Euclid's proof shows that this is true.

In Figure 7.4(b), a line is dropped from C to point K, this divides the larger square into two smaller rectangles $ABKJ$ and $KIBJ$. If we can show that the area of $ABKJ$ (the blue rectangle) is the same as the area of square $ACDE$ (the blue square) and the area of $KIBJ$ (the lilac rectange) is the same as the area of square $CBFG$ (the lilac square), then the proof of Pythagoras' Theorem will be complete.

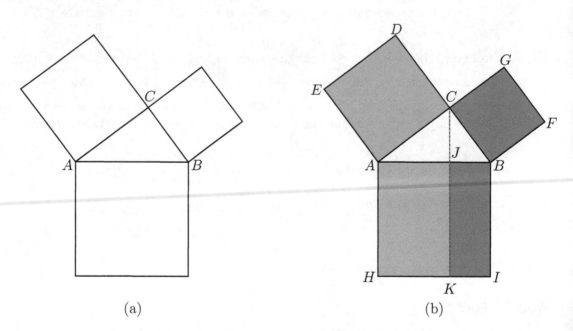

(a)   (b)

Figure 7.4: The first steps of Euclid's proof.

To do the above, we first draw lines BE and CH and create two new triangles ABE and AHC, as shown in Figure 7.5(a). Length $|EA|$ is the same as $|AC|$ and $|AH|$ is the same as $|AB|$ and $\angle EAB = 90° + \angle CAB = \angle CAH$. Thus, using (P1), ABE and AHC are congruent triangles and have the same area. Using (P3), we know that the area of EAB is half the area of rectangle $AHKJ$. Similarly, the area of EAB is half the area of square $ACDE$. As EAB and EAB are congruent, from (P2), their areas are the same and, hence,

$AHKJ$ has the same area as $ACDE$. Or, put more simply, the areas of the blue squares are the same.

In an almost identical manner, by drawing the lines AF and CI, as in Figure 7.5(b), and following the same logical steps, we can see that the areas of the two lilac squares are the same. The sum of the areas of the blue and lilac squares is this equal to the area of square $HIBA$.

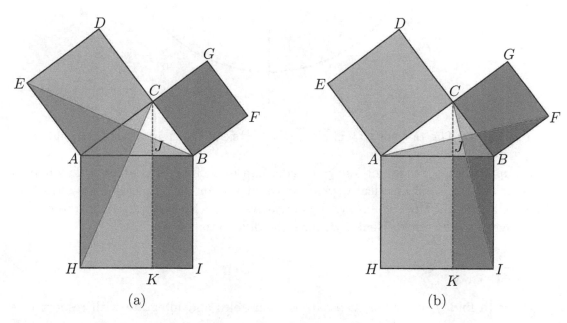

(a)   (b)

Figure 7.5: Congruent triangles can be used to show (a) the area of the blue square and rectangle are the same, (b) the area of the lilac square and rectangle are the same.

.3 Proof using Calculus

Pythagoras' theorem can also be proved using techniques from calculus and the direct link with the equation of a circle found.

Consider Figure 7.6, which shows a right angled triangle OAB, with $|OA| = a$, $|AB| = b$ and $|BA| = c$. Now consider a circle of radius c centred at O. Clearly point B must lie on the circle. The gradient of OB is $\frac{b}{a}$. Hence, the gradient of the tangent to the circle is $m = -\frac{a}{b}$.

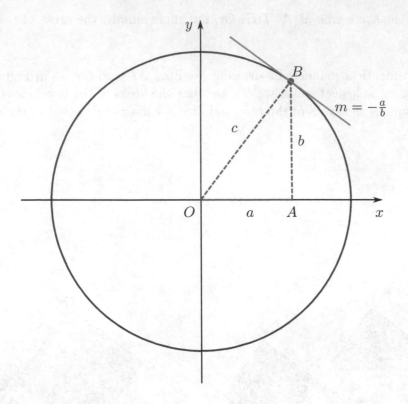

Figure 7.6: The tangent to a circle at the vertex of the right angled triangle OAB.

Now suppose that we do not know the equation of a circle, but assume that it is of the form $y = y(x)$. In an identical way to above, at a point (x, y) on the circle, the gradient of the tangent must be $-\frac{x}{y}$. Now, the derivative of y with respect to x is the gradient of the tangent to the circle. Hence, we have the differential equation

$$\frac{\mathrm{d}y}{\mathrm{d}x} = -\frac{x}{y}.$$

In order to find y, we simply separate the variables and integrate with respect to x, to obtain

$$y\frac{\mathrm{d}y}{\mathrm{d}x} = -x$$

$$\Rightarrow \quad \int y\frac{\mathrm{d}y}{\mathrm{d}x}\,\mathrm{d}x = -\int x\,\mathrm{d}x,$$

$$\Rightarrow \quad \int y\,\mathrm{d}y = -\int x\,\mathrm{d}x,$$

$$\Rightarrow \quad \frac{y^2}{2} = -\frac{x^2}{2} + k,$$

$$\Rightarrow \quad y^2 = -x^2 + K. \tag{7.1}$$

Where $K = 2k$ is a constant of integration. Now, as we are finding the equation of the circle, we know that at $x = 0$, $y = c$. Hence we can substitute these values into (7.1), to obtain

$$c^2 = K.$$

Hence, the equation for the circle is:

$$x^2 + y^2 = c^2.$$

As our original point (a, b) lies on this circle, it too follows that

$$a^2 + b^2 = c^2.$$

Proof Tip

When proving a theorem, it is important that we do not rely on results whose proof in turn relied on the theorem we are currently trying to prove: this is termed a *circular argument*.

In the above, we made use of two results. The first is that the tangent to a circle is perpendicular to the radius, the second is that if two lines with gradients m_1 and m_2 are perpendicular, then $m_1 m_2 = -1$. Both of these results can be proved using Pythagoras' theorem and we will see the latter's proof later in this chapter. If using Pythagoras' theorem was the only means of proving these results, then clearly we would not be able to use them to prove Pythagoras' theorem. Fortunately, both results can be proved by purely geometric means.

.4 Converse of Pythagoras' Theorem

Pythagoras' Theorem tells us that, if we have a right angled triangle, then the length of the sides satisfy the equation $a^2 + b^2 = c^2$, where c is the length of the hypotenuse. Is the converse true? *i.e.* if we have a triangle where the lengths of the sides satisfy $a^2 + b^2 = c^2$, is it necessarily right angled? The answer is yes, as we show below. In many proofs, it is useful to know that the converse is true, rather than Pythagoras' Theorem itself.

Theorem 7.1 — Converse of Pythagoras' Theorem

Suppose a triangle T has sides of length a, b and c and that $a^2 + b^2 = c^2$, then the triangle is right angled.

Proof:

The proof makes use of Pythagoras' theorem. Suppose that we have a right angled triangle $\hat{T}$ with shorter sides of length a and b, then, by Pythagoras' theorem, we must have that $a^2 + b^2 = x^2$, where x is the length of the hypotenuse. Rearranging gives

$$x = \pm\sqrt{a^2 + b^2}.$$

Of course, only the positive length $x = \sqrt{a^2 + b^2}$ is physically sensible. Now, we can similarly rearrange $a^2 + b^2 = c^2$, so that

$$c = \sqrt{a^2 + b^2}.$$

Hence, $c = x$. We, therefore, have two triangles with sides of the same length, hence, the triangles must be congruent. As $\hat{T}$ is right angled, it follows that T is also right angled.

7.2 Proofs that Require Pythagoras' Theorem

As mentioned in Chapter 1, one of the beautiful aspects of mathematics is that more and more complex theorems and proofs can be built using results that have come before. Indeed, in Euclid's proof of Pythagoras' theorem, the three results (P1-P3) were required. Many of the geometric results used in pre University mathematics can be proved using Pythagoras' theorem, or its converse. In this section we investigate some of these proofs.

7.2.1 Finding the Midpoint of a Line Segment

> **Formula 7.1 — The Midpoint of a Line Segment**
> The coordinates of the midpoint of the line segment joining the points $A(x_A, y_A)$ and $B(x_B, y_B)$ are
>
> $$\left(\frac{x_A + x_B}{2}, \frac{y_A + y_B}{2} \right).$$

Proof:

Pythagoras theorem can be used to show that the point C with coordinates $\left(\frac{x_A + x_B}{2}, \frac{y_A + y_B}{2} \right)$ has the property that $|AC| = |CB|$, *i.e.* C is equidistant from A and B. To do this, we simply make use of the formula for the distance between two points, which is just an application of Pythagoras' theorem. If $D(x_D, y_D)$ and $E(x_E, y_E)$ are two points, then the distance between these two points is

$$|DE| = \sqrt{(x_E - x_D)^2 + (y_E - y_D)^2}.$$

Using this formula, we directly show that $|AC| = |CB|$:

$$
\begin{aligned}
|AC| &= \sqrt{\left(\frac{x_A + x_B}{2} - x_A \right)^2 + \left(\frac{y_A + y_B}{2} - y_A \right)^2} \\
&= \sqrt{\left(\frac{x_B - x_A}{2} \right)^2 + \left(\frac{y_B - y_A}{2} \right)^2} \\
&= \sqrt{\left(x_B - \frac{x_A + x_B}{2} \right)^2 + \left(y_B - \frac{y_A + y_B}{2} \right)^2} \\
&= |CB|.
\end{aligned}
$$

7.2.2 Gradient Condition for Perpendicular Line Segments

> **Formula 7.2 — Gradient Condition for Perpendicular Line Segments**
> For the points $A(x_A, y_A)$, $B(x_B, y_B)$, $C(x_C, y_C)$ and $D(x_D, y_D)$, let the line segments AB and CD have well defined gradients m_1 and m_2, respectively. Then, AB and CD are perpendicular if and only if
>
> $$m_1 m_2 = -1. \tag{7.2}$$

Proof:

Without loss of generality, we consider the situation where the points B and D coincide, and so the line segments join each other at one end, as shown in Figure 7.7.

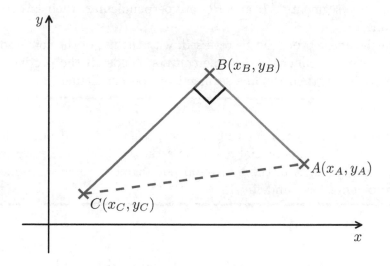

Figure 7.7: Line segment AB is perpendicular to line segment BC.

The gradients m_1, m_2 of the line segments AB, and BC respectively, are

$$m_1 = \frac{y_B - y_A}{x_B - x_A},$$
$$m_2 = \frac{y_C - y_B}{x_C - x_B}.$$

Now, since AB is perpendicular to BC, triangle ABC is right-angled and so Pythagoras' theorem can be applied:

$$|AB|^2 + |BC|^2 = |AC|^2. \tag{7.3}$$

Considering the left hand side of Equation (7.3),

$$|AB|^2 + |BC|^2 = (x_B - x_A)^2 + (y_B - y_A)^2 + (x_C - x_B)^2 + (y_B - y_C)^2. \tag{7.4}$$

Considering the right hand side of Equation (7.3),

$$
\begin{aligned}
|AC|^2 &= (x_C - x_A)^2 + (y_C - y_A)^2, \\
&= (x_C + x_B - x_B - x_A)^2 + (y_C + y_B - y_B - y_A)^2, \\
&= [(x_C - x_B) + (x_B - x_A)]^2 + [(y_C - y_B) + (y_B - y_A)]^2, \\
&= (x_C - x_B)^2 + (x_B - x_A)^2 + 2(x_C - x_B)(x_B - x_A) + (y_C - y_B)^2 \\
&\quad + (y_B - y_A)^2 + 2(y_C - y_B)(y_B - y_A).
\end{aligned}
\tag{7.5}
$$

Since Equation (7.3) holds,

$$0 = 2(x_C - x_B)(x_B - x_A) + 2(y_C - y_B)(y_B - y_A),$$
$$\Rightarrow \quad -2(y_C - y_B)(y_B - y_A) = 2(x_C - x_B)(x_B - x_A),$$
$$\Rightarrow \quad \frac{(y_C - y_B)(y_B - y_A)}{(x_C - x_B)(x_B - x_A)} = \frac{2}{-2},$$
$$\Rightarrow \quad \frac{(y_C - y_B)}{(x_C - x_B)} \frac{(y_B - y_A)}{(x_B - x_A)} = -1,$$
$$\Rightarrow \quad m_1 m_2 = -1.$$

Thus, if the two line segments AB and BC are perpendicular, their gradients multiply to give negative one.

All the steps in the above proof can be reversed, with the use of the converse of Pythagoras' theorem, and so we also conclude that the converse holds: if the product of gradients of two line segments is -1, then the line segments are perpendicular.

> **Remark**
> We notice that two line segments will also be perpendicular if one is horizontal and the other vertical. This is a special case where the gradient of the vertical segment is undefined and the gradient of the horizontal line segment equal to zero. The multiple of the two gradients is also undefined.

> **Proof Tip**
> The algebra in the above proof is rather heavy and hard to read. When writing it out by hand, it would be very likely that some mistakes could be made. We notice that the same quantities, such as $(x_C - x_B)$ and $(y_B - y_A)$, appear multiple times. We could use some new notation to write these more concisely, for example, x_{CB} could be used to represent $x_C - x_B$. In doing this, the potential for introducing mistakes into the algebra will be reduced.

> **Proof Tip**
> In deriving Equation (7.5), we have twice used a technique that can often be applied when trying to prove an assertion - the addition of zero. For example, $(x_C - x_A)$ became $(x_C + x_B - x_B - x_A)$ as $+x_B - x_B = 0$ and so adding this leaves the original expression unchanged.
> A similar technique makes use of the multiplicative identity to multiply a term by 1 in a clever way to enable progress to be made in a proof. Look out for this in other proofs.

7.2.3 Angle in a Semicircle (Thales' Theorem)

Thales, like Pythgoras, was an Ancient Greek philosopher. He gave his name to an important theorem, which we commonly refer to as the angle in a semicircle.

> **Interactive Activity 7.1 — Angle in a Semicircle**
> Use the Geogebra applet below in the digital book to investigate the angles in a triangle with one side a diameter and all three vertices on the circle.

> **Theorem 7.2 — Thales' Theorem (Angle in a semicircle)**
> If a triangle is constructed with one of its sides being the diameter of a circle and the opposite point lying on the circle, then the triangle is right angled, see Figure 7.8.

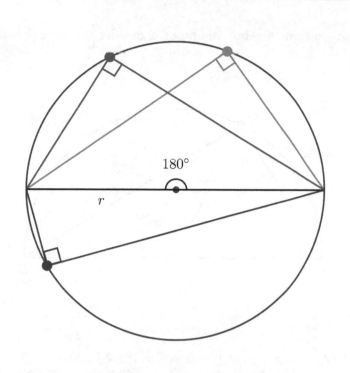

Figure 7.8: The triangle formed from the endpoints of the diameter of a circle and another point on the circumference of the circle is right angled.

Proof:

There are many possible proofs of this theorem, however, we make use of the general equation of a circle here, as it leads to an application of the converse of Pythagoras' Theorem. As it is easy to see that rotation and translation of the circle will make no difference to the result, we consider a circle centred at the origin for simplicity and without loss of generality. The equation of the circle is thus

$$x^2 + y^2 = r^2,$$

where r is the radius of the circle. Similarly, which diameter is picked is unimportant, hence we consider a triangle with vertices at $A = (-r, 0)$, $B = (r, 0)$ and $C = (x, y)$, where (x, y) satisfies $x^2 + y^2 = r^2$. We now calculate the length of each of the sides:

$$|AB|^2 = (2r)^2,$$
$$|BC|^2 = |(x, y) - (r, 0)|^2 = (x - r)^2 + y^2 = x^2 - 2rx + r^2 + y^2 = 2r^2 - 2rx,$$
$$|CA|^2 = |(-r, 0) - (x, y)|^2 = (-r - x)^2 + (-y)^2 = x^2 + 2rx + r^2 + y^2 = 2r^2 + 2rx.$$

Hence, $|BC|^2 + |CA|^2 = |AB|^2$ and, by the converse of Pythagoras' theorem, we have a right angled triangle.

Remark

For completeness, we show a geometric proof. Consider the figure shown below.

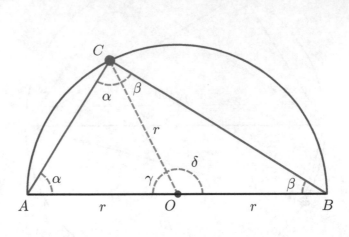

By joining point O to C, we form two isosceles triangles, both with common sides of length r. Using properties of triangles, we know that $\gamma = 180 - 2\alpha$ and $\delta = 180 - 2\beta$. We also know that $\gamma + \delta = 180$. combining these, we obtain

$$
\begin{aligned}
180 &= \gamma + \delta \\
&= 180 - 2\alpha + 180 - 2\beta \\
&= 360 - 2(\alpha + \beta), \\
\Rightarrow \quad 180 &= 2(\alpha + \beta), \\
\Rightarrow \quad 90 &= \alpha + \beta.
\end{aligned}
$$

The angle is $90°$ as required.

Exercise 7.1

State the converse of Thales' theorem and prove that this is also true.

7.2.4   Perpendicular Bisector of a Chord

Interactive Activity 7.2 — Perpendicular Bisector of a Chord

Use the Geogebra applet below in the digital book to investigate the perpendicular bisector of a chord of a circle.

Theorem 7.3 — Perpendicular Bisector of a Chord

Suppose that any chord of a circle is drawn and a line is constructed from the midpoint of the chord to the centre of the circle. This line bisects the chord at an angle of 90°, *i.e.* it is a perpendicular bisector of the chord. Two examples of this are shown in Figure 7.9.

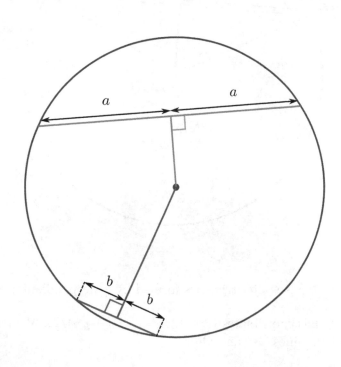

Figure 7.9: The perpendicular bisector of a chord of a circle passes through the centre of the circle.

Proof:

Once again, there are many possible proofs of Theorem 7.3, we make use of the converse of Pythagoras' theorem to prove it.

Without loss of generality, we consider a circle centred at the origin O and a chord passing through the points $A(x_A, y_A)$ and $B(x_B, y_B)$ that lie on the circle. This situation can be seen in Figure 7.10. Letting C be the midpoint of $\overrightarrow{AB}$, we will show that the triangle AOC is always right angled and hence the line OC is a perpendicular bisector of AB.

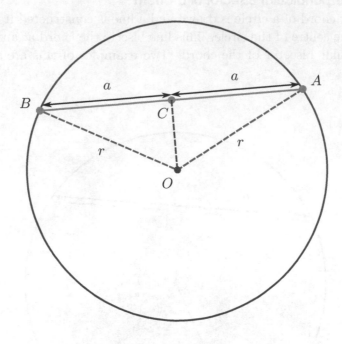

Figure 7.10: Two isosceles triangles formed by a perpendicular of a chord.

From Formula 7.1, C has coordinates $((x_A + x_B)/2, (y_A + y_B)/2)$. We now find the square of the magnitude of the lines AC and OC:

$$|AC|^2 = \left((x_A - x_B)^2 + (y_A - y_B)^2\right)/4,$$
$$|OC|^2 = \left((x_A + x_B)^2 + (y_A + y_B)^2\right)/4.$$

Adding these together, we find

$$\begin{aligned}
|AC|^2 + |OC|^2 &= \left((x_A - x_B)^2 + (y_A - y_B)^2\right)/4 + \left((x_A + x_B)^2 + (y_A + y_B)^2\right)/4 \\
&= \left(2(x_A^2 + y_A^2) + 2(x_B^2 + y_B^2)\right)/4 \\
&= \frac{4r^2}{4} \\
&= r^2.
\end{aligned}$$

Now, as $|OA| = r$, the converse of Pythagoras' theorem reveals that AOC must be a right angled triangle, as required.

Exercise 7.2

State the converse of Theorem 7.3 and prove that it is also true.

Remark

There are many alternative ways to prove Theorem 7.3. In the following, we use vectors to prove the result. With the same notation as before and referring back to Figure 7.10, we know that

$$\overrightarrow{AC} = ((x_A - x_B)/2, (y_A - y_B)/2),$$
$$\overrightarrow{OC} = ((x_A + x_B)/2, (y_A + y_B)/2).$$

Hence,

$$\begin{aligned}
\overrightarrow{AC} \cdot \overrightarrow{OC} &= ((x_A - x_B)(x_A + x_B) + (y_A - y_B)(y_A + y_B))/4 \\
&= (x_A^2 + y_A^2 - (x_B^2 + y_B^2))/4 \\
&= (r^2 - r^2)/4 \\
&= 0.
\end{aligned}$$

By properties of the dot product, the two vectors are therefore perpendicular to each other.

7.2.5   The Cosine Rule

The cosine rule relates the lengths of the sides of a triangle a, b, c and one of the angles of the triangle. It is essentially a generalisation of Pyhagoras' theorem and it is, therefore, not unexpected that the proof of the cosine rule can be shown using Pythagoras' theorem.

> **Theorem 7.4 — Cosine rule**
> For a triangle ABC
>
> $$c^2 = a^2 + b^2 - 2ab\cos(C), \tag{7.6}$$
>
> where a is the length of the side opposite apex A, *etc.*

Proof
We prove that the theorem holds when C is an acute angle.

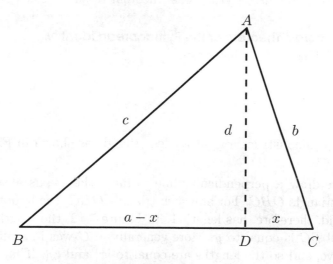

Figure 7.11: Triangle for use in the proof of the cosine rule.

Figure 7.11 shows a triangle ABC, split into two right angled triangles so that the line CD is perpendicular to AB.

By defining $x = |DC|$, we use Pythagoras' theorem to find d in both triangles ADC and DBC

$$d^2 = c^2 - (a - x)^2 \quad \text{and} \quad d^2 = b^2 - x^2.$$

Hence, noting that $x = c\cos(B)$, we have

$$
\begin{aligned}
c^2 - (a - x)^2 &= b^2 - x^2, \\
\Rightarrow \quad c^2 - a^2 + 2ax - x^2 &= b^2 - x^2, \\
\Rightarrow \quad c^2 &= a^2 + b^2 - 2ax, \\
\Rightarrow \quad c^2 &= a^2 + b^2 - 2ab\cos(C).
\end{aligned}
$$

We could use any of the angles as the apex of this triangle which leads to two other versions of Cosine Rule:

$$
\begin{aligned}
a^2 &= b^2 + c^2 - 2bc\cos(A), \\
b^2 &= a^2 + c^2 - 2ac\cos(B).
\end{aligned}
$$

Exercise 7.3

Construct a proof of Theorem 7.4 where angle C is obtuse.

7.2.6 The Pythagorean Identity in Trigonometry

Our final result of the chapter uses Pythagoras' theorem to show that

---

Theorem 7.5 — Pythagoras' Theorem or the Pythagorean Identity.
For any angle θ the following identity is true.

$$\sin^2(\theta) + \cos^2(\theta) \equiv 1.$$

Proof:
Given any angle θ we can construct a right angled triangle as shown in Figure 7.12.

From the point C, we draw a perpendicular line to meet the x-axis at a point B. This forms a right-angled triangle OBC. For any such triangle OBC, the hypotenuse OC is the radius of the circle and, therefore, has length 1. In Figure 7.12, the length of OB is equal to x and the length of BC is equal to y. More generally, if C was in a different quadrant, x or y may be negative, and so the lengths are equal to $|x|$ and $|y|$. That is,

$$|OB| = |x| = |\cos(\theta)| \quad \text{and} \quad |BC| = |y| = |\sin(\theta)|.$$

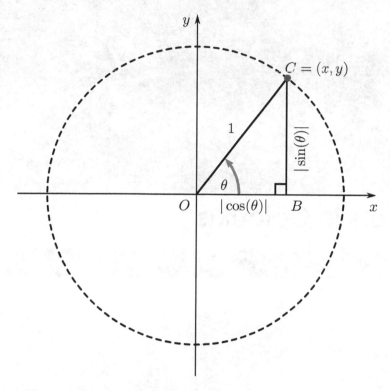

Figure 7.12: A right angled triangle in the unit circle.

We can apply Pythagoras' Theorem to this triangle to give us the result:

$$|\sin(\theta)|^2 + |\cos(\theta)|^2 \equiv 1,$$
$$\Rightarrow \quad \sin^2(\theta) + \cos^2(\theta) \equiv 1.$$

Remark

We note the use of the $\equiv$ symbol, rather than $=$, in Theorem 7.5. $\equiv$ means equivalent. This is a better notation in this case, because the result holds for all θ, it is not an equation where we are trying to find a θ that makes the equality true.

8. Proofs in Calculus

Calculus is fundamental to many areas of applied mathematics, physics, engineering and other sciences. As such, the results need to be grounded with rigorous proof. In this chapter, we present some of the key ideas of differential and integral calculus.

8.1 Differential Calculus

Before we consider the important proofs associated with differential calculus, we must recall the definition of a derivative.

Definition 8.1 — The Derivative
The derivative of the curve $y = f(x)$ is denoted either as $f'(x)$ or $\frac{dy}{dx}$. We define $f'(x)$ to be

$$f'(x) = \lim_{h \to 0} \frac{f(x+h) - f(x)}{h}. \tag{8.1}$$

Remark
In Definition 8.1, $\lim_{h \to 0}$ means "in the limit as h tends to 0". A limit has a strict definition itself, but we omit this at this time. Importantly, we cannot evaluate

$$\frac{f(x+h) - f(x)}{h} \tag{8.2}$$

at $h = 0$, as this would involve dividing by zero. However, we can let h become arbitrarily close to 0 and, as this happens, the expression (8.2) approaches a limiting value. It is this value that is defined to be the derivative at x.

Limits have a number of useful properties, which we will exploit later. For two functions

$f(h)$ and $g(h)$ and $c \in \mathbb{R}$, we have:

$$\lim_{h \to 0} (f(h) + g(h)) = \lim_{h \to 0} f(h) + \lim_{h \to 0} g(h),$$

$$\lim_{h \to 0} cf(h) = c \lim_{h \to 0} f(h).$$

These show that *taking limits* is a linear operation. These two formulae are part of a larger set of rules known as *the algebra of limits*.

.1 Differentiation from First Principles

The definition of a derivative can be used to directly find the gradient of simple functions. This process is known as *differentiation from first principles*. We give a number of examples of its use below.

Theorem 8.2 — Derivative of x

The derivative of $f(x) = x$ is 1 for all $x \in \mathbb{R}$.

Proof:
We apply the definition of a derivative directly to $f(x) = x$:

$$\begin{aligned}
f'(x) &= \lim_{h \to 0} \frac{f(x+h) - f(x)}{h} \\
&= \lim_{h \to 0} \frac{(x+h) - x}{h} \\
&= \lim_{h \to 0} \frac{h}{h} \\
&= \lim_{h \to 0} 1 \\
&= 1.
\end{aligned}$$

Theorem 8.3 — Derivative of x^2

The derivative of $f(x) = x^2$ is $2x$ for all $x \in \mathbb{R}$.

Proof:
We apply the definition of a derivative directly to $f(x) = x^2$ to find

$$\begin{aligned}
f'(x) &= \lim_{h \to 0} \frac{f(x+h) - f(x)}{h} \\
&= \lim_{h \to 0} \left(\frac{(x+h)^2 - x^2}{h} \right) \\
&= \lim_{h \to 0} \left(\frac{x^2 + 2xh + h^2 - x^2}{h} \right) \\
&= \lim_{h \to 0} (2x + h) \\
&= 2x.
\end{aligned}$$

Differentiation from first principles can be used to find the derivative of $f(x) = x^n$, where n is any rational number. The proof relies on the binomial expansion.

Theorem 8.4 — The Derivative of $f(x) = x^n$

Let $f(x) = x^n$, where $n \in \mathbb{N}_0$, then

$$f'(x) = \frac{\mathrm{d}f(x)}{\mathrm{d}x} = nx^{n-1}.$$

Proof:

We first prove the cases $n = 0$ and $n = 1$, before employing the binomial theorem to expand $(x + h)^n$ for all other positive n.

Case 1: $n = 0$

In this case $f(x) = x^0 = 1$. By Definition 8.1,

$$\begin{aligned}
\frac{\mathrm{d}f(x)}{\mathrm{d}x} &= \lim_{h \to 0} \frac{f(x+h) - f(x)}{h} \\
&= \lim_{h \to 0} \frac{1 - 1}{h} \\
&= \lim_{h \to 0} \frac{0}{h} \\
&= 0.
\end{aligned}$$

Here, $nx^{n-1} = 0x^{-1} = 0$, and so our assertion is true for $n = 0$.

Case 2: $n = 1$

For $n = 1$, $f(x) = x^1 = x$. Using Definition 8.1 again

$$\begin{aligned}
\frac{\mathrm{d}f(x)}{\mathrm{d}x} &= \lim_{h \to 0} \frac{f(x+h) - f(x)}{h} \\
&= \lim_{h \to 0} \frac{x + h - x}{h} \\
&= \lim_{h \to 0} \frac{h}{h} \\
&= 1.
\end{aligned}$$

Now, for $n = 1$, $nx^{n-1} = 1x^{1-1} = x^0 = 1$, proving the statement for $n = 1$.

Case 3: $n > 1$

We first expand $(x + h)^n$ using the binomial theorem,

$$(x + h)^n = x^n + nx^{n-1}h + \frac{n(n-1)}{2}x^{n-2}h^2 + \ldots + h^n.$$

Then we use Definition 8.1 to find the derivative of $f(x) = x^n$ from first principles:

$$
\begin{aligned}
\frac{\mathrm{d}f(x)}{\mathrm{d}x} &= \lim_{h \to 0} \frac{f(x+h) - f(x)}{h} \\
&= \lim_{h \to 0} \frac{(x+h)^n - x^n}{h} \\
&= \lim_{h \to 0} \frac{x^n + nx^{n-1}h + \frac{n(n-1)}{2}x^{n-2}h^2 + \ldots + h^n - x^n}{h} \\
&= \lim_{h \to 0} \frac{nx^{n-1}h + \frac{n(n-1)}{2}x^{n-2}h^2 + \ldots + h^n}{h} \\
&= \lim_{h \to 0} \frac{h\left(nx^{n-1} + \frac{n(n-1)}{2}x^{n-2}h + \ldots h^{n-1}\right)}{h} \\
&= \lim_{h \to 0} \left(nx^{n-1} + \frac{n(n-1)}{2}x^{n-2}h + \ldots h^{n-1}\right) \\
&= nx^{n-1}.
\end{aligned}
$$

Hence, as we have shown the assertion to be true for $n = 0$, $n = 1$ and integers greater than 1 it is true for all positive integers.

Remark

Later we will show that the same result holds for all $n \in \mathbb{Q}$.

In the following example we show how differentiation from first principles can be used to differentiate more complicated functions.

Example 8.1

Show, from first principles, that the derivative of $f(x) = \sin(x)$ is $f'(x) = \cos(x)$.

Solution:

From the definition we have

$$
f'(x) = \lim_{\delta x \to 0} \frac{\sin(x + \delta x) - \sin(x)}{\delta x}.
$$

Applying the addition formula for $\sin(A + B)$ with $A = x$ and $B = \delta x$ gives:

$$
f'(x) = \lim_{\delta x \to 0} \frac{\sin(x)\cos(\delta x) + \cos(x)\sin(\delta x) - \sin(x)}{\delta x}.
$$

Then, using the small angle approximations, we obtain

$$
\begin{aligned}
f'(x) &= \lim_{\delta x \to 0} \frac{\sin(x)\left(1 - \frac{1}{2}(\delta x)^2\right) + \cos(x)(\delta x) - \sin(x)}{\delta x} \\
&= \lim_{\delta x \to 0} \frac{\sin(x) - \frac{1}{2}(\delta x)^2 \sin(x) + \delta x \cos(x) - \sin(x)}{\delta x} \\
&= \lim_{\delta x \to 0} \frac{\delta x \cos(x) - \frac{1}{2}(\delta x)^2 \sin(x)}{\delta x} \\
&= \lim_{\delta x \to 0} \cos(x) - \frac{1}{2}\delta x \sin(x).
\end{aligned}
$$

As $\delta x \to 0$, $\cos(x) - \frac{1}{2}\delta x \sin(x) \to \cos(x)$ and so $f'(x) = \cos(x)$.

Remark

In the example above, we note that x is measured in radians, otherwise the small angle approximations do not hold. If x was measured in degrees, the differentiation result is not true.

Exercise 8.1

Show from first principles that if $f(x) = \cos(x)$, then $f'(x) = -\sin(x)$.

8.1.2 Further Properties of Differentiation

So far, we have mainly only considered the derivatives of simple monomials, the results proved in the next theorem, when combined with the previous results, allow us to find the derivative of any polynomial and provide a first step into finding derivatives without needing to carry out differentiation from first principles.

Theorem 8.5 — Properties of Differentiation

Let $f(x)$ and $g(x)$ be continuous, differentiable functions and $c \in \mathbb{R}$, then,

(a)

$$\frac{\mathrm{d}}{\mathrm{d}x}\left[f(x) + g(x)\right] = \frac{\mathrm{d}}{\mathrm{d}x}\left[f(x)\right] + \frac{\mathrm{d}}{\mathrm{d}x}\left[g(x)\right], \tag{8.3}$$

(b)

$$\frac{\mathrm{d}}{\mathrm{d}x}\left[f(x) - g(x)\right] = \frac{\mathrm{d}}{\mathrm{d}x}\left[f(x)\right] - \frac{\mathrm{d}}{\mathrm{d}x}\left[g(x)\right], \tag{8.4}$$

(c)

$$\frac{\mathrm{d}}{\mathrm{d}x}\left[cf(x)\right] = c\frac{\mathrm{d}}{\mathrm{d}x}\left[f(x)\right]. \tag{8.5}$$

Proof:

(a) Using Definition 8.1,

$$\frac{d}{dx}[f(x) + g(x)] = (f(x) + g(x))'$$

$$= \lim_{h \to 0} \left(\frac{f(x+h) + g(x+h) - (f(x) + g(x))}{h} \right)$$

$$= \lim_{h \to 0} \left(\frac{f(x+h) - f(x) + g(x+h) - g(x)}{h} \right)$$

$$= \lim_{h \to 0} \left(\frac{f(x+h) - f(x)}{h} \right) + \lim_{h \to 0} \left(\frac{g(x+h) - g(x)}{h} \right)$$

$$= \frac{d}{dx}[f(x)] + \frac{d}{dx}[g(x)].$$

In the above we have used the linearity properties of limits.

(b) This is analogous to part (a) with a negative sign instead of the positive sign.

(c) Again, using Definition 8.1,

$$\frac{d}{dx}[cf(x)] = \lim_{h \to 0} \frac{cf(x+h) - cf(x)}{h}$$

$$= \lim_{h \to 0} c \left(\frac{f(x+h) - f(x)}{h} \right)$$

$$= c \lim_{h \to 0} \frac{f(x+h) - f(x)}{h} \qquad \text{(using properties of limits)}$$

$$= c \frac{d}{dx}[f(x)].$$

Corollary 8.6

Let $f_1(x), \ldots, f_n(x)$ be differentiable functions and $c_1, \ldots, c_n \in \mathbb{R}$, if

$$f(x) = \sum_{i=1}^{n} c_i f_i(x),$$

then

$$\frac{df(x)}{dx} = \sum_{i=1}^{n} c_i \frac{df_i(x)}{dx}.$$

Proof:

This result follows using the results of Theorem 8.5 and an induction argument. Let $P(n)$ be the statement:

"The derivative of $f(x) = \sum_{i=1}^{n} c_i f_i(x)$ is $\frac{df(x)}{dx} = \sum_{i=1}^{n} c_i \frac{df_i(x)}{dx}$."

Step 1:

Let us show that the case $P(2)$ is true. We have

$$\sum_{i=1}^{2} c_i f_i(x) = c_1 f_1(x) + c2 f_2(x),$$

$$\Rightarrow \quad \frac{d}{dx}[c_1 f_1 + c_2 f_2] = \frac{d}{dx}[c_1 f_1] + \frac{d}{dx}[c_2 f_2] \quad \text{(Using (8.3))}$$

$$= c_1 \frac{df_1}{dx} + c_2 \frac{df_2}{dx}. \quad \text{(Using (8.5))}$$

Step 2:

We now assume that $P(k)$ is true for some $k \geq 2$.

Step 3:

To show that $P(k+1)$ is true, we write

$$\sum_{i=1}^{k+1} c_i f_i(x) = \underbrace{\sum_{i=1}^{k} c_i f_i(x)}_{g(x)} + c_{k+1} f_{k+1}(x),$$

$$\Rightarrow \quad \frac{d}{dx}\left[\sum_{i=1}^{k+1} c_i f_i\right] = \frac{d}{dx}[g + c_{k+1} f_{k+1}]$$

$$= \frac{dg}{dx} + \frac{d}{dx}[c_{k+1} f_{k+1}] \quad \text{(Using (8.3))}$$

$$= \frac{dg}{dx} + c_{k+1} \frac{df_{k+1}}{dx} \quad \text{(Using (8.4))}$$

$$= \sum_{i=1}^{k} c_i \frac{df_i}{dx} + c_{k+1} \frac{df_{k+1}}{dx} \quad \text{(Using } P(k)\text{)}$$

$$= \sum_{i=1}^{k+1} c_i \frac{df_i}{dx}.$$

Step 4:

By the principle of mathematical induction, $P(n)$ must hold for all $n \geq 2$.

With the results of Theorem 8.5 and Corollary 8.6, we can extend previous results for monomials to general polynomials.

Theorem 8.7 — The Derivative of $f(x) = ax^n$

Let $f(x) = ax^n$, where $n \in \mathbb{N}_0$, then

$$f'(x) = \frac{df(x)}{dx} = anx^{n-1} \qquad (8.6)$$

Proof:

It suffices to apply the third part of Theorem 8.5 to the result from Theorem 8.4.

Theorem 8.8 — Derivatives of Polynomials

Let $f(x)$ be an nth order polynomial so that $f(x) = \sum_{i=0}^{n} a_i x^i$, where $n \in \mathbb{N}$, $a_i \in \mathbb{R}$ and $a_n \neq 0$, then

$$f'(x) = \frac{\mathrm{d}f(x)}{\mathrm{d}x} = \sum_{i=1}^{n} i a_i x^{i-1}. \tag{8.7}$$

Proof:

The proof of this result follows on application of Corollary 8.6 and Theorem 8.7.

$$f'(x) = \frac{\mathrm{d}}{\mathrm{d}x}\left[\sum_{i=0}^{n} a_i x^i\right]$$

$$= \sum_{i=0}^{n} \frac{\mathrm{d}}{\mathrm{d}x}[a_i x^i]$$

$$= \sum_{i=0}^{n} i a_i x^{i-1}$$

$$= \sum_{i=1}^{n} i a_i x^{i-1}.$$

Proof Tip

We notice that each theorem in this chapter builds upon earlier theorems and the results become more and more complex and general.

Formula 8.1 — Product Rule

For a function defined as a product $y(x) = u(x)v(x)$, where u and v are both differentiable functions of x, the derivative can be expressed as:

$$\frac{\mathrm{d}y}{\mathrm{d}x} = v\frac{\mathrm{d}u}{\mathrm{d}x} + u\frac{\mathrm{d}v}{\mathrm{d}x}. \tag{8.8}$$

Proof:

For a function $y(x) = u(x) \cdot v(x)$

$$y'(x) = \lim_{h \to 0} \frac{u(x+h) \cdot v(x+h) - u(x) \cdot v(x)}{h}$$

$$= \lim_{h \to 0} \frac{u(x+h) \cdot v(x+h) - u(x) \cdot v(x) + u(x) \cdot v(x+h) - u(x) \cdot v(x+h)}{h}$$

$$= \lim_{h \to 0} \frac{v(x+h)[u(x+h) - u(x)] + u(x)[v(x+h) - v(x)]}{h}$$

$$= \lim_{h \to 0} v(x+h) \cdot \lim_{h \to 0} \frac{u(x+h) - u(x)}{h} + \lim_{h \to 0} u(x) \cdot \lim_{h \to 0} \frac{v(x+h) - v(x)}{h}$$

$$= v(x) \cdot u'(x) + u(x) \cdot v'(x).$$

Proof Tip

In the seond line of the previous proof, we once again employed the common technique of addition of a zero term.

Exercise 8.2

Q1. Show that a function of the form $y = u \cdot v \cdot w$ has derivative

$$\frac{dy}{dx} = uv\frac{dw}{dx} + u\frac{dv}{dx}w + \frac{du}{dx}vw.$$

Q2. Let $g(x) = \prod_{i=1}^{n} f_i(x)$ for differentiable functions $f_i(x)$, $i = 1, \ldots, n$. Suggest a formula for the derivative of $g(x)$ in terms of $f_i(x)$ and their derivatives. Use induction to prove the formula is correct.

Recall, $\prod_{i=1}^{n} f_i(x)$ is shorthand for $f_1(x) \times f_2(x) \times \ldots \times f_n(x)$.

Formula 8.2 — Quotient Rule

For a function defined as a quotient such as $y = \frac{u}{v}$, where u and $v \neq 0$ are both functions of x, the derivative can be expressed as:

$$\frac{dy}{dx} = \frac{v\frac{du}{dx} - u\frac{dv}{dx}}{v^2}. \tag{8.9}$$

Exercise 8.3

Q1. Prove the quotient rule directly from first principles using techniques similar to those used for the product rule.

Q2. Use the quotient rule to find the derivatives of:
(a) $\tan(x)$;
(b) $\sec(x)$;
(c) $\operatorname{cosec}(x)$.

Remark

Although we now have a method for functions written as a quotient it is also possible to use the product rule for these types of functions. Suppose we need to differentiate $y = \frac{2x^3}{\sqrt{3x^2-5}}$. We can apply the quotient rule with $u = 2x^3$ and $v = (3x^2 - 5)^{\frac{1}{2}}$. Alternatively, we could apply the product rule with $u = 2x^3$ and $v = (3x^2 - 5)^{-\frac{1}{2}}$.

Formula 8.3 — The Chain Rule

Let $y = y(u)$, and u be a function of x. Assuming y is differentiable with respect to u and u is differentiable with respect to x, then y can be differentiated with respect to x according to the following formula:

$$\frac{dy}{dx} = \frac{dy}{du} \times \frac{du}{dx}. \tag{8.10}$$

> **Remark**
>
> When using the chain rule it is important to understand where to evaluate the derivatives. When evaluating at a point x, we expand on the equation shown in (8.10) to make this clear:
>
> $$\frac{dy}{dx}(x) = \frac{dy}{du}(u(x)) \times \frac{du}{dx}(x).$$

Proof:

The proof of the chain rule requires some careful manipulation of limits. Here we present the general idea of the proof, without going into too much technical detail. We begin with the definition of the derivative of $y = y(u(x))$ and work to extract a term involving the derivative of u by multiplying through by 1.

$$
\begin{aligned}
\frac{dy}{dx}(x) &= \lim_{h \to 0} \frac{y(u(x+h)) - y(u(x))}{h} \\
&= \lim_{h \to 0} 1 \cdot \frac{y(u(x+h)) - y(u(x))}{h} \\
&= \lim_{h \to 0} \frac{u(x+h) - u(x)}{u(x+h) - u(x)} \cdot \frac{y(u(x+h)) - y(u(x))}{h} \quad \text{(Multiplication by 1)} \\
&= \lim_{h \to 0} \frac{y(u(x+h)) - y(u(x))}{u(x+h) - u(x)} \cdot \frac{u(x+h) - u(x)}{h} \\
&= \lim_{h \to 0} \frac{y(u(x+h)) - y(u(x))}{u(x+h) - u(x)} \cdot \lim_{h \to 0} \frac{u(x+h) - u(x)}{h} \quad \text{(Algebra of Limits)} \\
&= \lim_{h \to 0} \frac{y(u(x+h)) - y(u(x))}{u(x+h) - u(x)} \cdot \frac{du}{dx}(x).
\end{aligned}
$$

If we assume y and u are continuous functions with $u(x) \neq 0$, it can be shown that

$$\frac{y(u(x+h)) - y(u(x))}{u(x+h) - u(x)} \to \frac{dy}{dx}(u(x)) \text{ as } h \to 0,$$

although this step does require a more rigorous use of limits, which would take us on an unnecessary diversion at this level. With this result, we immediately have that

$$\frac{dy}{dx}(x) = \frac{dy}{du}(u(x)) \times \frac{du}{dx}(x).$$

Exercise 8.4

Q1. Use the chain rule to show that, if $v(x)$ is a differentiable function and $v(x) \neq 0$,

$$\left(\frac{1}{v}\right)'(x) = -\frac{v'(x)}{v^2(x)}.$$

Q2. Use the result from Q1 to prove the quotient rule using an application of the product rule.

Q3. Suppose θ is an angle measured in degrees. Use the chain rule to find the derivatives, with respect to θ, of
   (a) $\sin(\theta)$;
   (b) $\cos(\theta)$;

> (c) $\tan(\theta)$.
>
> [Hint: recall the transformation that takes radians to degrees.]

Theorem 8.9 — The Derivative of $f(x) = x^n$, for $n \in \mathbb{Q}$

Let $f(x) = x^n$, where $n \in \mathbb{Q}$, then

$$f'(x) = \frac{\mathrm{d}f(x)}{\mathrm{d}x} = nx^{n-1}. \tag{8.11}$$

We have already seen how to prove the same result when $n \in \mathbb{N}$. To prove the result for rational n, we make use of the chain rule and the product rule. We build the final result up in steps and use the fact that, if $n \in \mathbb{Q}$, then $n = \frac{k}{m}$ for $k \in \mathbb{Z}$ and $m \in \mathbb{N}$.

Step 1: $n = \frac{1}{m}$:
Let

$$f(x) = x^{1/m}, \tag{8.12}$$

for some $m \in \mathbb{N}$, so that $n = 1/m$. We can raise both sides of (8.12) to the power m to obtain

$$[f(x)]^m = [x^{1/m}]^m = x. \tag{8.13}$$

Now, we can differentiate both sides of (8.13), the left hand side using the chain rule and Theorem 8.4 and the right hand side just using Theorem 8.4, to give:

$$\frac{\mathrm{d}}{\mathrm{d}x}[f(x)]^m = \frac{\mathrm{d}x}{\mathrm{d}x},$$
$$\Rightarrow \quad mf'(x)[f(x)]^{m-1} = 1,$$
$$\Rightarrow \quad mf'(x)(x^{1/m})^{m-1} = 1,$$
$$\Rightarrow \quad mf'(x)x^{1-1/m} = 1,$$
$$\Rightarrow \quad f'(x) = \frac{1}{m}x^{1/m-1}.$$

Hence, for $n = 1/m$,

$$f'(x) = \frac{\mathrm{d}f(x)}{\mathrm{d}x} = nx^{n-1}.$$

Step 2: $n = \frac{k}{m}$:
Now let

$$f(x) = x^{k/m}, \tag{8.14}$$

for some k and $m \in \mathbb{N}$, so that $n = k/m > 0$, then laws of indices gives

$$f(x) = (x^{1/m})^k, \tag{8.15}$$

We can use the chain rule directly and the result from Step 1 and Theorem 8.4 as follows:

$$\frac{\mathrm{d}f(x)}{\mathrm{d}x} = \frac{\mathrm{d}}{\mathrm{d}x}[(x^{1/m})^k]$$

$$= k(x^{1/m})^{k-1} \cdot \frac{1}{m}x^{1/m-1}$$

$$= \frac{k}{m}x^{k/m-1}.$$

Hence, for $n = k/m$, with k and $m \in \mathbb{N}$, we have

$$f'(x) = \frac{\mathrm{d}f(x)}{\mathrm{d}x} = nx^{n-1}.$$

Step 3: $n = -\frac{k}{m}$:
Now let

$$f(x) = x^{-k/m}, \tag{8.16}$$

for some k and $m \in \mathbb{N}$, so that $n = -k/m < 0$. Then, we can multiply (8.16) on both sides by $x^{k/m}$ to obtain

$$f(x)x^{k/m} = 1. \tag{8.17}$$

We can then differentiate both sides of (8.17), using the product rule on the left hand side together with the results from Step 2 to obtain

$$\frac{\mathrm{d}}{\mathrm{d}x}[f(x)x^{k/m}] = \frac{\mathrm{d}1}{\mathrm{d}x},$$

$$\Rightarrow \quad f'(x)x^{k/m} + f(x)\frac{k}{m}x^{k/m-1} = 0,$$

$$\Rightarrow \quad f'(x)x^{k/m} + x^{-k/m}\frac{k}{m}x^{k/m-1} = 0,$$

$$\Rightarrow \quad f'(x)x^{k/m} + \frac{k}{m}x^{-1} = 0,$$

$$\Rightarrow \quad f' = -\frac{\frac{k}{m}x^{-1}}{x^{k/m}}$$

$$= -\frac{k}{m}x^{-k/m-1}.$$

Hence, for $n = -k/m$, with k and $m \in \mathbb{N}$, we have

$$f'(x) = \frac{\mathrm{d}f(x)}{\mathrm{d}x} = nx^{n-1}.$$

Finally, we note that the case $n = 0$ has already been covered in Theorem 8.4, hence the proof is complete for all $n \in \mathbb{Q}$.

Proof Tip

The proof of Theorem 8.4 for $n \in \mathbb{N}_0$ used the binomial expansion theorem. Initially, the authors thought that, to prove Theorem 8.9, the rational version of the binomial expansion would be required. However, this did not seem possible and they sought an alternative method to prove Theorem 8.9. It is natural that initial ideas for proofs may not work, do not be disheartened, but think of alternative means to prove a result.

Remark

Theorem 8.9 actually holds for all $n \in \mathbb{R}$, but the proof of this is beyond the scope of this book.

8.2 Integral Calculus

The other main branch of calculus is *integral calculus*, where the motivation is to find the area A between a curve C, defined by a function $f(x)$, and the x-axis, between the limits $x = a$ and $x = b$, as shown in Figure 8.1.

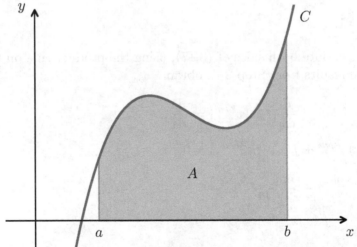

Figure 8.1: The area A under a general curve C between the limits $x = a$ and $x = b$.

Before we prove some of the main theorems of integral calculus, we recall some notation.

Definition 8.10 — The Definite Integral

The *definite integral* of a continuous function $f(x)$ between the limits $x = a$ and $x = b$ is the area between $f(x)$ and the x-axis and is denoted by

$$\int_a^b f(x)\,\mathrm{d}x.$$

> **Definition 8.11 — Antiderivative**
> F is called an antiderivative of a function f if it is differentiable and $F'(x) = f(x)$ for all x.

2.1 Fundamental Theorems of Calculus

One of the most important results in Calculus is the Fundamental Theorem of Calculus that links the ideas of integration and differentiation and shows how the integral of a function is related to its antiderivative.

> **Theorem 8.12 — Fundamental Theorem of Calculus Part I**
> Let $f(x)$ be a continuous function defined on the closed interval $[a, b]$ and let
>
> $$F(x) = \int_a^x f(x)\, dx,$$
>
> where $x \in [a, b]$. Then, $F(x)$ is differentiable on (a, b) and
>
> $$F'(x) = f(x).$$

Proof:
Suppose curve C in Figure 8.2 is defined by $f(x)$. Let

$$F(x) = \int_a^x f(x)\, dx$$

be the area under the curve between the limits a and an arbitrary point x that lies between a and b. From Section 8.1, we know that the derivative of F can be calculated as

$$F'(x) = \lim_{h \to 0} \frac{F(x+h) - F(x)}{h}. \tag{8.18}$$

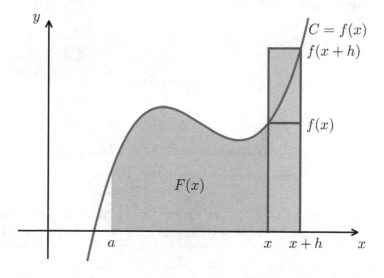

Figure 8.2: The area $F(x)$ under the curve $C = f(x)$ between the limits $x = a$ and an arbitrary point x. The rectangles between x and $x + h$ used in the derivation below are also shown.

Now, $F(x+h) - F(x)$ is the area under the curve $f(x)$ between x and $x+h$. We construct two possible rectangles to approximate this area. Bearing in mind Figure 8.2, the first approximation is given by the rectangle with area $f(x) \cdot h$, while the second is given by the rectangle with area $f(x+h) \cdot h$. As $h \to 0$ and because $f(x)$ is continuous, there will come a point where $f(x+h) \geq f(x)$, or $f(x+h) \leq f(x)$. Without loss of generality, let us assume $f(x+h) \geq f(x)$.

Calculating the areas of the corresponding rectangles gives

$$f(x) \cdot h \leq F(x+h) - F(x) \leq f(x+h) \cdot h.$$

Therefore,

$$f(x) \leq \frac{F(x+h) - F(x)}{h} \leq f(x+h).$$

We know that, as $h \to 0$, $x + h \to x$ and $f(x+h) \to f(x)$, because f is continuous. Therefore,

$$\lim_{h \to 0} \frac{F(x+h) - F(x)}{h} = f(x).$$

Substituting into (8.18) we obtain:

$$F'(x) = f(x). \tag{8.19}$$

In other words, $f(x)$ is equal to the derivative of the area under f between the limits a and x.

The following Corollary gives us a methodology for evaluating a definite integral.

Corollary 8.13

If F' is continuous on $[a, b]$, then

$$\int_a^b F'(x)\, dx = F(b) - F(a).$$

In other words, if F is the integral of f on $[a, b]$, then

$$\int_a^b f(x)\, dx = F(b) - F(a).$$

Proof Tip

A *corollary* is a minor result that follows simply from a more major theorem. A *lemma* is a minor result that is used on the way to proving a more major theorem.

Exercise 8.5

Prove Corollary 8.13.

Remark

Corollary 8.13 is a slightly weaker version of the theorem known as the *Fundamental Theorem of Calculus Part II*.

8.2 Further Properties of Integration

We now briefly discuss some important properties of integration.

Theorem 8.14

Let $f(x)$ and $g(x)$ be continuous functions on the interval $[a, b]$, then

$$\int_a^b f(x) + g(x)\,\mathrm{d}x = \int_a^b f(x)\,\mathrm{d}x + \int_a^b g(x)\,\mathrm{d}x \qquad (8.20)$$

In other words, if F is the integral of f and G is the integral of g:

$$\int_a^b f(x) + g(x)\,\mathrm{d}x = F(b) - F(a) + G(b) - G(a).$$

Similarly,

$$\int_a^b f(x) - g(x)\,\mathrm{d}x = \int_a^b f(x)\,\mathrm{d}x + \int_b^a g(x)\,\mathrm{d}x \qquad (8.21)$$

In other words, if F is the integral of f and G is the integral of g:

$$\int_a^b f(x) - g(x)\,\mathrm{d}x = F(b) - F(a) + G(a) - G(b).$$

Exercise 8.6

Prove Theorem 8.14.

As we have seen, the Fundamental Theorem of Calculus reveals that integration is the reverse of differentiation. Hence, we can use the reverse of properties of differentiation to help us perform integrations. Two important methods are shown below.

Theorem 8.15 — Integration by Parts

Suppose $u(x)$ and $v(x)$ are two differentiable functions for $x \in (a, b)$, then

$$\int_a^b u\frac{\mathrm{d}v}{\mathrm{d}x}\,\mathrm{d}x = [uv]_a^b - \int_a^b v\frac{\mathrm{d}u}{\mathrm{d}x}\,\mathrm{d}x.$$

Theorem 8.16 — Integration by Substitution

Suppose $f = f(u)$ is an integrable function and u is a differentiable function of x, for

$x \in [a, b]$. Then f is a function of x and it can be integrated using the formula:

$$\int_a^b f(u(x)) \frac{\mathrm{d}u}{\mathrm{d}x} \, \mathrm{d}x = \int_{u(a)}^{u(b)} f(u) \, \mathrm{d}u.$$

Exercise 8.7

Q1. Prove Theorem 8.15 by making use of the Fundamental Theorems of Calculus and the product rule.

Q2. Prove Theorem 8.16 by making use of the Fundamental Theorems of Calculus and the chain rule.

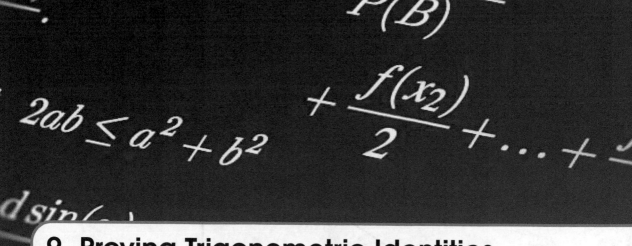

Consider the trigonometric equation,

$$\frac{\tan(\theta) + \sin(\theta)}{\tan(\theta)} = \frac{2}{3}.$$

(9.1)

It is not immediately obvious how this can be solved. With equations like these, however, it is often worth attempting to simplify the left hand side. To do this we must employ known trigonometric identities, or prove that the left hand side is equivalent to another function which results in an equation which is simpler to solve.

In this chapter we first prove some standard trigonometric results, and then apply these results to prove some further trigonometric identities.

.1 The Standard Results

In the following we assume familiarity with the trigonometric functions below, defined for θ measured in either radians or degrees:

$$\sin(\theta)$$

$$\cos(\theta)$$

$$\tan(\theta) = \frac{\sin(\theta)}{\cos(\theta)}$$

$$\sec(\theta) = \frac{1}{\cos(\theta)}$$

$$\csc(\theta) = \frac{1}{\sin(\theta)}$$

$$\cot(\theta) = \frac{1}{\tan(\theta)} = \frac{\cos(\theta)}{\sin(\theta)}$$

9.1.1 The Pythagorean Trigonometric Identities

Consider the triangle below,

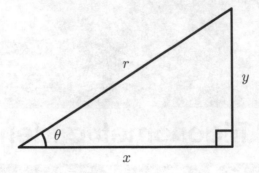

Figure 9.1: A general right angled triangle.

Due to Pythagoras' Theorem,

$$x^2 + y^2 = r^2,$$

$$\Rightarrow \quad \frac{x^2}{r^2} + \frac{y^2}{r^2} = \frac{r^2}{r^2},$$

$$\Rightarrow \quad \frac{x^2}{r^2} + \frac{y^2}{r^2} = 1,$$

$$\Rightarrow \quad \left(\frac{x}{r}\right)^2 + \left(\frac{y}{r}\right)^2 = 1$$

Since $\sin(\theta) = \frac{y}{r}$ and $\cos(\theta) = \frac{x}{r}$ we have that,

$$\cos^2(\theta) + \sin^2(\theta) = 1.$$

This identity is more usually presented in the following form.

Formula 9.1

$$\sin^2(\theta) + \cos^2(\theta) \equiv 1. \tag{9.2}$$

From Formula 9.1 we can derive two further identities based on an application of Pythagoras' Theorem.

Dividing both sides of identity (9.2) by $\cos^2(\theta)$ we have,

$$\sin^2(\theta) + \cos^2(\theta) \equiv 1$$

$$\Rightarrow \quad \frac{\sin^2(\theta)}{\cos^2(\theta)} + \frac{\cos^2(\theta)}{\cos^2(\theta)} \equiv \frac{1}{\cos^2(\theta)}$$

$$\Rightarrow \quad \left(\frac{\sin(\theta)}{\cos(\theta)}\right)^2 + 1 \equiv \left(\frac{1}{\cos(\theta)}\right)^2$$

$$\Rightarrow \quad \tan^2(\theta) + 1 \equiv \sec^2(\theta).$$

Similarly, dividing both sides of identity (9.2) by $\sin^2(\theta)$ we have,

$$\sin^2(\theta) + \cos^2(\theta) \equiv 1$$
$$\Rightarrow \quad \frac{\sin^2(\theta)}{\sin^2(\theta)} + \frac{\cos^2(\theta)}{\sin^2(\theta)} \equiv \frac{1}{\sin^2(\theta)}$$
$$\Rightarrow \quad 1 + \left(\frac{\cos(\theta)}{\sin(\theta)}\right)^2 \equiv \left(\frac{1}{\sin(\theta)}\right)^2$$
$$\Rightarrow \quad 1 + \cot^2(\theta) \equiv \operatorname{cosec}^2(\theta).$$

Rearranging the above identities into their common form we have the three Pythagorean trigonometric identities.

Formulae 9.2 — Pythagorean Trigonometric Identities

$$\sin^2(\theta) + \cos^2(\theta) \equiv 1, \tag{9.3}$$
$$\sec^2(\theta) \equiv 1 + \tan^2(\theta), \tag{9.4}$$
$$\operatorname{cosec}^2(\theta) \equiv 1 + \cot^2(\theta). \tag{9.5}$$

.2 The Addition Formulae

Consider Figure 9.2.

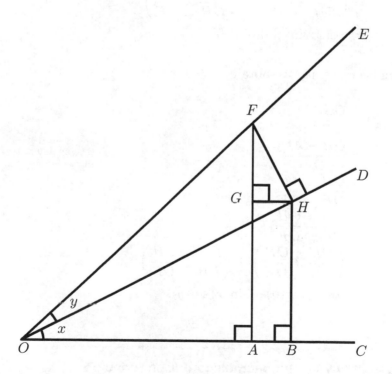

Figure 9.2: A construction for proving the addition formulae.

In Figure 9.2 angles x, y and $x + y$ are all acute. FA and HB are both perpendicular to OC, FH is perpendicular to OD and GH is perpendicular to FA.

As GH is parallel to OC,

$$\angle GHO = \angle HOC = x.$$

Since FH is perpendicular to OD, $\angle FHO = 90°$, and so $\angle FHG = 90° - x$. As the angles in a triangle sum to $180°$,

$$
\begin{aligned}
& \angle FGH + \angle FHG + \angle HFG = 180° \\
\Rightarrow \quad & 90° + (90° - x) + \angle HFG = 180° \\
\Rightarrow \quad & 180° - x = \angle HFG = 180° \\
\Rightarrow \quad & \angle HFG = x.
\end{aligned}
$$

By definitiion,

$$
\begin{aligned}
\sin(x + y) &= \frac{FA}{OF} \\
&= \frac{FG + GA}{OF} \\
&= \frac{FG}{OF} + \frac{GA}{OF} \\
&= \frac{FG}{OF} + \frac{HB}{OF} \\
&= \left(\frac{FG}{FH} \times \frac{FH}{OF} \right) + \left(\frac{HB}{OH} \times \frac{OH}{OF} \right) \\
&= \cos(x) \sin(y) + \sin(x) \cos(y).
\end{aligned}
$$

Following similar reasoning we obtain,

$$
\begin{aligned}
\cos(x + y) &= \frac{OA}{OF} \\
&= \frac{OB - AB}{OF} \\
&= \frac{OB}{OF} - \frac{AB}{OF} \\
&= \frac{OB}{OF} - \frac{GH}{OF} \\
&= \left(\frac{OB}{OH} \times \frac{OH}{OF} \right) - \left(\frac{GH}{FH} \times \frac{FH}{OF} \right) \\
&= \cos(x) \cos(y) - \sin(x) \sin(y).
\end{aligned}
$$

Interactive Activity 9.1 — Trigonometric Addition Formulae
Using the geogebra applet below explore the addition identities.

From the identities for $\sin(x+y)$ and $\cos(x+y)$ we can obtain an identity for $\tan(x+y)$.

$$\begin{aligned}
\tan(x+y) &= \frac{\sin(x+y)}{\cos(x+y)} \\
&= \frac{\sin(x)\cos(y) + \cos(x)\sin(y)}{\cos(x)\cos(y) - \sin(x)\sin(y)} \\
&= \frac{\dfrac{\sin(x)\cos(y)}{\cos(x)\cos(y)} + \dfrac{\cos(x)\sin(y)}{\cos(x)\cos(y)}}{\dfrac{\cos(x)\cos(y)}{\cos(x)\cos(y)} - \dfrac{\sin(x)\sin(y)}{\cos(x)\cos(y}} \\
&= \frac{\tan(x) + \tan(y)}{1 - \tan(x)\tan(y)},
\end{aligned}$$

where we have used $\tan(x) = \frac{\sin(x)}{\cos(x)}$ in the last line.

We now have three identities, we can obtain a further three by replacing y with $-y$ in the corresponding identities. Hence,

$$\begin{aligned}
\sin(x-y) &= \sin(x)\cos(-y) + \cos(x)\sin(-y) \\
&= \sin(x)\cos(y) - \cos(x)\sin(y)
\end{aligned}$$

$$\begin{aligned}
\cos(x-y) &= \cos(x)\cos(-y) - \sin(x)\sin(-y) \\
&= \cos(x)\cos(y) + \sin(x)\sin(y)
\end{aligned}$$

$$\begin{aligned}
\tan(x-y) &= \frac{\tan(x) + \tan(-y)}{1 - \tan(x)\tan(-y)} \\
&= \frac{\tan(x) - \tan(y)}{1 - \tan(x)(-\tan(y))} \\
&= \frac{\tan(x) - \tan(y)}{1 + \tan(x)(\tan(y)}
\end{aligned}$$

Proof Tip
The above derivations have depended on x and y being acute, however these identities hold true for all values of x and y.

The 6 identities derived above are collectively known as the addition formulae.

Formulae 9.3 — The Addition Formulae

$$\sin(x + y) \equiv \sin(x)\cos(y) + \cos(x)\sin(y), \tag{9.6}$$
$$\sin(x - y) \equiv \sin(x)\cos(y) - \cos(x)\sin(y), \tag{9.7}$$
$$\cos(x + y) \equiv \cos(x)\cos(y) - \sin(x)\sin(y), \tag{9.8}$$
$$\cos(x - y) \equiv \cos(x)\cos(y) + \sin(x)\sin(y), \tag{9.9}$$
$$\tan(x + y) \equiv \frac{\tan(x) + \tan(y)}{1 - \tan(x)\tan(y)}, \tag{9.10}$$
$$\tan(x - y) \equiv \frac{\tan(x) - \tan(y)}{1 + \tan(x)(\tan(y)}. \tag{9.11}$$

9.1.3 Double and Half Angle Formulae

From the addition formula for $\sin(x + y)$ we can derive an identity known as the double angle formulae for sine. Let $y = x$ in Identity (9.6), then

$$\sin(x + x) \equiv \sin(x)\cos(x) + \cos(x)\sin(x)$$
$$\Rightarrow \quad \sin(2x) = 2\sin(x)\cos(x)$$

Similarly, with the use of the Pythagorean Identity (9.5) we can obtain three double angle identities for cosine.

$$\cos(x + x) = \cos(x)\cos(x) - \sin(x)\sin(x)$$
$$\Rightarrow \quad \cos(2x) = \cos^2(x) - \sin^2(x)$$

Now, since $\cos^2(x) = 1 - \sin^2(x)$,

$$\cos(2x) = \cos^2(x) - \sin^2(x)$$
$$= 1 - \sin^2(x) - \sin^2(x)$$
$$= 1 - 2\sin^2(x).$$

Of course, we also know that $\sin^2(x) = 1 - \cos^2(x)$, and so,

$$\cos(2x) = \cos^2(x) - \sin^2(x)$$
$$= \cos^2(x) - (1 - \cos^2(x))$$
$$= 2\cos^2(x) - 1$$

Working similarly for the tangent function we have,

$$\tan(x + x) = \frac{\tan(x) + \tan(x)}{1 - \tan(x)\tan(x)}$$
$$\Rightarrow \quad \tan(2x) = \frac{2\tan(x)}{1 - \tan^2(x)}$$

Formulae 9.4 — Double Angle Formulae

$$\sin(2x) \equiv 2\sin(x)\cos(x), \tag{9.12}$$
$$\cos(2x) \equiv \cos^2(x) - \sin^2(x) \tag{9.13}$$
$$\equiv 1 - 2\sin^2(x) \tag{9.14}$$
$$\equiv 2\cos^2(x) - 1, \tag{9.15}$$
$$\tan(2x) \equiv \frac{2\tan(x)}{1 - \tan^2(x)}. \tag{9.16}$$

From the double angle formulae for cosine, letting $x = \frac{1}{2}\theta$, we can derive the half angle formulae for sine and cosine.

$$\cos(2x) = 1 - 2\sin^2(x)$$
$$\Rightarrow \quad 2\sin^2(x) = 1 - \cos(2x)$$
$$\Rightarrow \quad \sin^2(x) = \frac{1}{2}(1 - \cos(2x))$$
$$\Rightarrow \quad \sin^2\left(\frac{1}{2}\theta\right) = \frac{1}{2}(1 - \cos(\theta))$$

Similarly,

$$\cos(2x) = 2\cos^2(x) - 1$$
$$\Rightarrow \quad 2\cos^2(x) = 1 + \cos(2x)$$
$$\Rightarrow \quad \cos^2(x) = \frac{1}{2}(\cos(2x) + 1)$$
$$\Rightarrow \quad \cos^2\left(\frac{1}{2}\theta\right) = \frac{1}{2}(\cos(\theta) + 1)$$

Formulae 9.5 — Half Angle Formulae

$$\sin^2\left(\frac{1}{2}\theta\right) \equiv \frac{1}{2}(1 - \cos(\theta)), \tag{9.17}$$
$$\cos^2\left(\frac{1}{2}\theta\right) \equiv \frac{1}{2}(\cos(\theta) + 1). \tag{9.18}$$

.4 Sum and Product Formulae

Using, once again, the addition identities for sine we know that

$$\sin(x + y) \equiv \sin(x)\cos(y) + \cos(x)\sin(y), \quad ①$$
$$\sin(x - y) \equiv \sin(x)\cos(y) - \cos(x)\sin(y). \quad ②$$

Adding ① to ② we obtain,

$$2\sin(x)\cos(y) \equiv \sin(x + y) + \sin(x - y),$$

and, subtracting ② from ① we have,

$$2\cos(x)\sin(y) \equiv \sin(x+y) - \sin(x-y)$$

Similarly, using the addition identities for cosine,

$$\cos(x+y) \equiv \cos(x)\cos(y) - \sin(x)\sin(y), \qquad ③$$
$$\cos(x-y) \equiv \cos(x)\cos(y) + \sin(x)\sin(y). \qquad ④$$

Taking the sum of ③ and ④ we have,

$$2\cos(x)\cos(y) \equiv \cos(x+y) + \cos(x-y),$$

and considering the difference of ④ and ③,

$$2\sin(x)\sin(y) \equiv \cos(x-y) - \cos(x+y).$$

Letting $x + y = A$ and $x - y = B$ we can re-write these in an alternative form. Notice the following,

$$A + B = (x+y) + (x-y)$$
$$= 2x,$$
$$\Rightarrow \qquad x = \frac{A+B}{2},$$
$$A - B = (x+y) - (x-y)$$
$$= 2y,$$
$$\Rightarrow \qquad y = \frac{A-B}{2}.$$

Using these with the sum-product formulae we have derived above we have,

$$2\sin(x)\cos(y) \equiv \sin(x+y) + \sin(x-y)$$
$$\Rightarrow \quad \sin(A) + \sin(B) \equiv 2\sin\left(\frac{A+B}{2}\right)\cos\left(\frac{A-B}{2}\right),$$
$$2\cos(x)\sin(y) \equiv \sin(x+y) - \sin(x-y)$$
$$\Rightarrow \quad \sin(A) - \sin(B) \equiv 2\cos\left(\frac{A+B}{2}\right)\sin\left(\frac{A-B}{2}\right),$$
$$2\cos(x)\cos(y) \equiv \cos(x+y) + \cos(x-y)$$
$$\Rightarrow \quad \cos(A) + \cos(B) \equiv 2\cos\left(\frac{A+B}{2}\right)\cos\left(\frac{A-B}{2}\right),$$
$$2\sin(x)\sin(y) \equiv \cos(x-y) - \cos(x+y)$$
$$\Rightarrow \quad \cos(B) - \cos(A) \equiv 2\sin\left(\frac{A+B}{2}\right)\sin\left(\frac{A-B}{2}\right).$$

These eight identities are together known as the sum and product formulae.

Formulae 9.6 — Sum and Product Formulae

$$2\sin(x)\cos(y) \equiv \sin(x+y) + \sin(x-y), \tag{9.19}$$
$$2\cos(x)\sin(y) \equiv \sin(x+y) - \sin(x-y), \tag{9.20}$$
$$2\cos(x)\cos(y) \equiv \cos(x+y) + \cos(x-y), \tag{9.21}$$
$$2\sin(x)\sin(y) \equiv \cos(x-y) - \cos(x+y), \tag{9.22}$$
$$\sin(A) + \sin(B) \equiv 2\sin\left(\frac{A+B}{2}\right)\cos\left(\frac{A-B}{2}\right), \tag{9.23}$$
$$\sin(A) - \sin(B) \equiv 2\cos\left(\frac{A+B}{2}\right)\sin\left(\frac{A-B}{2}\right), \tag{9.24}$$
$$\cos(A) + \cos(B) \equiv 2\cos\left(\frac{A+B}{2}\right)\cos\left(\frac{A-B}{2}\right), \tag{9.25}$$
$$\cos(B) - \cos(A) \equiv 2\sin\left(\frac{A+B}{2}\right)\sin\left(\frac{A-B}{2}\right) \tag{9.26}$$

.5 The Small Angle Approximations

The small angle approximations provide low order polynomial approximations of the trigonometric functions $\sin(x)$, $\cos(x)$ and $\tan(x)$. These are only suitable for very small angles x. Using the definition of the trigonometric ratios for a right angle triangle we can geometrically derive the small angle approximations.

Consider the right angled triangle ABC shown in Figure 9.3, then, by trigonometry, the perpendicular height, h, can be calculated in the following two ways:

$$h = d\tan(\theta) \quad \text{and} \quad h = r\sin(\theta).$$

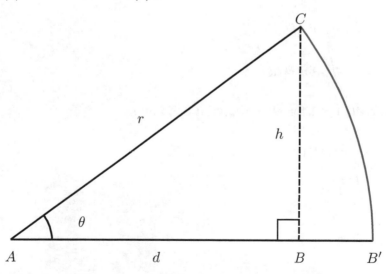

Figure 9.3: Derivation of the small angle approximations.

As the angle θ becomes close to zero, then the lengths r and d become approximately equal. In addition, the height h becomes close to the length of the circular arc joining B' to C, which can be calculated as $r\theta$ (provided the angle is given in radians). Hence, we have

$$h = r\sin(\theta) \approx r\theta \approx r\tan(\theta),$$

dividing by r leads to the approximations

$$\sin(\theta) \approx \tan(\theta) \approx \theta.$$

In order to obtain an approximation for $y = \cos(\theta)$, we make use of the double angle formula

$$\cos(2x) = 1 - 2\sin^2(x),$$

with $x = \frac{\theta}{2}$ and apply the small angle approximation for $\sin(x)$. Thus yielding,

$$\cos(\theta) = 1 - 2\sin^2\left(\frac{\theta}{2}\right)$$

$$\approx 1 - 2\left(\frac{\theta}{2}\right)^2$$

$$= 1 - \frac{\theta^2}{2}.$$

Remark

Supposing we know the standard calculus relations between $\sin(x)$ and $\cos(x)$; namely that

$$\frac{d}{dx}\sin(x) = \cos(x) \quad \text{and} \quad \frac{d}{dx}\cos(x) = \sin(x),$$

we can provide an alternative derivation of the small angle approximations. We also assume that $x \geq 0$ since, for negative x, we can make use of the identities $\sin(-x) = -\sin(x)$ and $\cos(-x) = \cos(x)$. By the Fundamental Theorem of Calculus,

$$\sin(x) = \int_0^x \cos(t)\,dt, \tag{9.27}$$

$$\cos(x) = 1 - \int_0^x \sin(t)\,dt. \tag{9.28}$$

The function $\cos(x) < 1$ for all $x \neq 2n\pi, n \in \mathbb{Z}$ and so

$$\sin(x) = \int_0^x \cos(t)\,dt$$

$$< \int_0^x 1\,dt$$

$$= x.$$

Similarly,

$$\cos(x) = 1 - \int_0^x \sin(t)\,dt$$

$$> 1 - \int_0^x t\,dt$$

$$= 1 - \frac{1}{2}x^2. \tag{9.29}$$

Using (9.27) and (9.29) we find that

$$\sin(x) = \int_0^x \cos(t)\,dt$$

$$> \int_0^x 1 - \frac{1}{2}t^2\,dt$$

$$= x - \frac{1}{6}x^3. \tag{9.30}$$

Hence, we have the following bounds for $y = \sin(x)$:

$$x - \frac{1}{6}x^3 < \sin(x) < x. \tag{9.31}$$

Now, if we use (9.28) together with (9.30) we have that,

$$\cos(x) = 1 - \int_0^x \sin(t)\,dt$$

$$< 1 - \int_0^x t - \frac{t^3}{6}\,dt$$

$$= 1 - \frac{1}{2}x^2 + \frac{1}{24}x^4. \tag{9.32}$$

Hence, we also have the following bounds on $y = \cos(x)$:

$$1 - \frac{1}{2}x^2 < \cos(x) < 1 - \frac{1}{2}x^2 + \frac{1}{24}x^5. \tag{9.33}$$

From (9.31) and (9.33) we can obtain small angle approximations and also note that our approximation for $\sin(x)$ will be an over approximation and that our approximation for $\cos(x)$ will be an under approximation.

Note, if this was the only way of deriving the small angle approximations we could not use them to differentiate $\sin(x)$ and $\cos(x)$ from first principles.

2   Applying the Standard Identities

With the standard identities that we have derived previously we can prove other identities or rearrange equations into a solvable form, as the next few examples will show. When trying to prove an identity of the form $A = B$, where A and B are expressions we need to perform some manipulations and use the identities from Section 9.1 to show that the left hand side is equal to the right hand side. In general, our argument will follow one fo the forms below.

- Simplify or rewrite A until we get B. This will give an argument which looks something like:

$$A = \cdots \qquad \text{because} \ldots$$
$$= \cdots \qquad \text{using} \ldots$$
$$= \cdots \qquad \text{using} \ldots$$
$$= B.$$

- The other way around: simplify or rewrite B until we get A.
- Simplify or rewrite A until we get some expression C, and then do the same with B. Then deduce that $A = B$. The argument will look something like the following:

$$
\begin{aligned}
A &= \cdots && \text{because} \dots \\
  &= \cdots && \text{using} \dots \\
  &= \cdots && \text{using} \dots \\
  &= C. \\
B &= \cdots && \text{because} \dots \\
  &= \cdots && \text{using} \dots \\
  &= \cdots && \text{using} \dots \\
  &= C.
\end{aligned}
$$

It follows that $A = B$.

- Show that $A - B = 0$, and deduce that $A = B$. It is sometimes easier to subtract two expressions and manipulate the difference, for example if one or both expressions involve fractions.
- Show that $A/B = 1$, and deduce that $A = B$. If both A and B involve fractions, it can make sense to divide them and show that the quotient equals one.
- Show that $A = B$ implies something else by manipulating both sides. Repeat this until we end up with something "obviously" true, such as $0 = 0$. As long as each line of implication is a two-way implication (as we learnt about in the Year 1 book, section 7.1.2), we can reverse the argument to deduce that $A = B$. When arguing in this way, it is vital to use the appropriate implication signs, otherwise the argument is invalid. Such an argument might look something like the following:

$$
\begin{aligned}
& A = B \\
\Leftrightarrow \quad & \cdots = \cdots && \text{dividing by} \dots \\
\Leftrightarrow \quad & \cdots = \cdots && \text{adding} \dots \\
\Leftrightarrow \quad & \cdots = \cdots && \text{using the identity} \dots \\
\Leftrightarrow \quad & \cdots = \cdots && \text{using the identity} \dots
\end{aligned}
$$

and the last statement is true because $\dots$, therefore the identity $A = B$ is true.

Remark

There are many identities in trigonometry for which one or both sides are not defined for some values of x. For example, $\tan(x) = \frac{\sin(x)}{\cos(x)}$ is true for all x with $x \neq (n + \frac{1}{2})\pi$ for any $n \in \mathbb{Z}$, but neither side is defined when $x = (n + \frac{1}{2})\pi$ for some integer n. When we state an identity, we will not usually specify the values of x for which it is valid, but assume that we mean that both sides are equal for all values of x for which both sides are defined.

We shall see most of the above forms of argument in the following examples.

Example 9.1

Prove that

$$(\sin(x) + \cos(x))(\sin(x) - \cos(x)) \equiv 2\sin^2(x) - 1$$

Solution:

We expand the left hand side of the identity given above.

$$(\sin(x) + \cos(x))(\sin(x) - \cos(x)) = \sin^2(x) - \sin(x)\cos(x) + \cos(x)\sin(x) - \cos^2(x)$$
$$= \sin^2(x) - \cos^2(x)$$
$$= \sin^2(x) - (1 - \sin^2(x))$$
$$= 2\sin^2(x) - 1.$$

Since this is true for all x we have proven the identity.

Proof Tip

At line two in the above we looked for which identity to apply. Applying 9.5 to the $-\cos^2(x)$ term led to the result. It should be clear that applying 9.5 to the $\sin^2(x)$ term would lead to another equivalent form of the left hand side, namely,

$$(\sin(x) + \cos(x))(\sin(x) - \cos(x)) \equiv 1 - 2\cos^2(x).$$

Example 9.2

Prove that

$$\cos^2(x) + 3\sin^2(x) \equiv 3 - 2\cos^2(x)$$

Solution:

For this example we shall start with the right hand side and use identity (9.5) twice to show that it is equal to the left hand side.

$$RHS = 3 - 2\cos^2(x)$$
$$\equiv 3 - 2(1 - \sin^2(x))$$
$$\equiv 1 + 2\sin^2(x)$$
$$\equiv (\cos^2(x) + \sin^2(x)) + 2\sin^2(x)$$
$$\equiv \cos^2(x) + 3\sin^2(x)$$
$$\equiv LHS.$$

Proof Tip

For Example 9.2 it could be argued that it is more elegant to start with the left hand side and do the following:

$$LHS = \cos^2(x) + 3\sin^2(x)$$
$$\equiv \cos^2(x) + 3(1 - \cos^2(x))$$
$$\equiv 3 - 2\cos^2(x)$$
$$\equiv RHS.$$

However, either approach is acceptable.

Example 9.3

Simplify the expression $\sin^4(\theta) + \sin^2(\theta)\cos^2(\theta)$.

Solution:

We use (9.5).

$$\begin{aligned}
\sin^4(\theta) + \sin^2(\theta)\cos^2(\theta) &= \sin^4(\theta) + \sin^2(\theta)\left(1 - \sin^2(\theta)\right), \\
&= \sin^4(\theta) + \sin^2(\theta) - \sin^4(\theta), \\
&= \sin^2(\theta).
\end{aligned}$$

Example 9.4

Prove the identity

$$\sec^2(x) + \operatorname{cosec}^2(x) = \sec^2(x)\operatorname{cosec}^2(x).$$

Solution:

It is not obvious at first glance how the two sides are related or which identity which might use to get started. We could therefore try writing each side in terms of $\sin(x)$ and $\cos(x)$, and see what happens:

$$\begin{aligned}
\sec^2(x) + \operatorname{cosec}^2(x) &= \frac{1}{\cos^2(x)} + \frac{1}{\sin^2(x)} \\
&= \frac{\sin^2(x)}{\sin^2(x)\cos^2(x)} + \frac{\cos^2(x)}{\sin^2(x)\cos^2(x)} \quad \text{(common denominator)} \\
&= \frac{\sin^2(x) + \cos^2(x)}{\sin^2(x)\cos^2(x)} \\
&= \frac{1}{\sin^2(x)\cos^2(x)} \quad \text{(Pythagorean identity)}
\end{aligned}$$

while

$$\begin{aligned}
\sec^2(x)\operatorname{cosec}^2(x) &= \frac{1}{\cos^2(x)} \times \frac{1}{\sin^2(x)} \\
&= \frac{1}{\sin^2(x)\cos^2(x)}.
\end{aligned}$$

Therefore $\sec^2(x) + \operatorname{cosec}^2(x) = \sec^2(x)\operatorname{cosec}^2(x)$.

Proof Tip

In Example 9.4 we used the approach of rewriting both A and B as the same third expression C.

Example 9.5

Prove the identity

$$\frac{\sec(x) - \operatorname{cosec}(x)}{\tan(x) - \cot(x)} = \frac{\tan(x) + \cot(x)}{\sec(x) + \operatorname{cosec}(x)}.$$

Solution:

We have fractions on both sides here, so it might make sense to try to remove or otherwise simplify the fractions. Here are three different approaches.

Approach 1: Manipulate the identity

We have

$$\frac{\sec(x) - \operatorname{cosec}(x)}{\tan(x) - \cot(x)} = \frac{\tan(x) + \cot(x)}{\sec(x) + \operatorname{cosec}(x)}$$

$$\Leftrightarrow \quad (\sec(x) - \operatorname{cosec}(x))(\sec(x) + \operatorname{cosec}(x)) =$$
$$(\tan(x) + \cot(x))(\tan(x) - \cot(x)) \quad (1)$$

$$\Leftrightarrow \qquad \sec^2(x) - \operatorname{cosec}^2(x) = \tan^2(x) - \cot^2(x) \quad (2)$$

$$\Leftrightarrow \qquad \sec^2(x) - \tan^2(x) = \operatorname{cosec}^2(x) - \cot^2(x) \quad (3)$$

$$\Leftrightarrow \qquad\qquad\qquad 1 = 1,$$

which is clearly true. Therefore the original identity is true.

Approach 2: Divide the sides of the identity

We have

$$\frac{\sec(x) - \operatorname{cosec}(x)}{\tan(x) - \cot(x)} \Big/ \frac{\tan(x) + \cot(x)}{\sec(x) + \operatorname{cosec}(x)}$$

$$= \frac{(\sec(x) - \operatorname{cosec}(x))(\sec(x) + \operatorname{cosec}(x))}{(\tan(x) + \cot(x))(\tan(x) - \cot(x))} \quad (1)$$

$$= \frac{\sec^2(x) - \operatorname{cosec}^2(x)}{\tan^2(x) - \cot^2(x)} \quad (2)$$

$$= \frac{(1 + \tan^2(x)) - (1 + \cot^2(x))}{\tan^2(x) - \cot^2(x)} \quad (3)$$

$$= \frac{\tan^2(x) - \cot^2(x)}{\tan^2(x) - \cot^2(x)} \quad (4)$$

$$= 1. \quad (5)$$

Therefore $\dfrac{\sec(x) - \operatorname{cosec}(x)}{\tan(x) - \cot(x)} = \dfrac{\tan(x) + \cot(x)}{\sec(x) + \operatorname{cosec}(x)}$ as required.

Approach 3: Simplify each fraction

We note that the denominator on the left-hand side, $\tan(x) - \cot(x)$, can be written as

$\dfrac{\sin(x)}{\cos(x)} - \dfrac{\cos(x)}{\sin(x)}$, so were we to write it as a single fraction, it would have a denominator of $\sin(x)\cos(x)$. Similarly, $\sec(x) + \text{cosec}(x)$, written in terms of $\sin(x)$ and $\cos(x)$, would have a denominator of $\sin(x)\cos(x)$. This suggests that we can simplify both fractions by multiplying their numerators and denominators by $\sin(x)\cos(x)$. We have:

$$\frac{\sec(x) - \text{cosec}(x)}{\tan(x) - \cot(x)} = \frac{\sec(x) - \text{cosec}(x)}{\tan(x) - \cot(x)} \times \frac{\sin(x)\cos(x)}{\sin(x)\cos(x)}$$

$$= \frac{\sin(x) - \cos(x)}{\sin^2(x) - \cos^2(x)}$$

$$= \frac{\sin(x) - \cos(x)}{(\sin(x) + \cos(x))(\sin(x) - \cos(x))}$$

$$= \frac{1}{\sin(x) + \cos(x)}$$

$$\frac{\tan(x) + \cot(x)}{\sec(x) + \text{cosec}(x)} = \frac{\tan(x) + \cot(x)}{\sec(x) + \text{cosec}(x)} \times \frac{\sin(x)\cos(x)}{\sin(x)\cos(x)}$$

$$= \frac{\sin^2(x) + \cos^2(x)}{\sin(x) + \cos(x)}$$

$$= \frac{1}{\sin(x) + \cos(x)}.$$

These are identical, and hence the original identity is true.

Proof Tip

We make some comments on each of the approaches shown in Example 9.5

Approach 1:

On line (1) (which has been split over two lines because it is so long), we multiplied both sides by the product of the denominators to clear the fractions; on line (2), we simplified the expressions by the difference of two squares identity, and on line (3), we used two Pythagorean identities.

Despite these being dual implications, working from bottom to top is far from obvious.

Approach 2:

After line (2), it was not clear what to do next, so we decided to try something: with the presence of $\sec^2(x)$ and $\text{cosec}^2\,x$, the Pythagorean identities seemed a sensible way to go. If we had applied them to both the numerator and denominator, we would have obtained:

$$\frac{(1 + \tan^2(x)) - (1 + \cot^2(x))}{(\sec^2(x) - 1) - (\text{cosec}^2(x) - 1)} = \frac{\tan^2(x) - \cot^2(x)}{\sec^2(x) - \text{cosec}^2(x)}.$$

At this point, it would be clear that we've done too much, as we now have the reciprocal of what we had on line (2), which is not that helpful. But we would see that if we only manipulated the numerator, we would end up with everything cancelling, as it does on line (4).

Approach 3: This is the approach that the authors feel is the most natural.

Example 9.6

Given that $x = a\cot(\theta)$, simplify the expression $\dfrac{x}{a^2 + x^2}$.

Solution:

We simply substitute for x in the expression, and then use identities to simplify:

$$\frac{x}{a^2 + x^2} = \frac{a\cot(\theta)}{a^2 + a^2\cot^2(\theta)}$$

$$= \frac{a\cot(\theta)}{a^2\operatorname{cosec}^2(\theta)}$$

$$= \frac{\cos(\theta)/\sin(\theta)}{a/\sin^2(\theta)}$$

$$= \frac{1}{a}\sin(\theta)\cos(\theta).$$

We can now return to the solution of the equation in the introduction of this chapter.

Example 9.7

Solve, finding all solutions in the range $0° \le \theta \le 360°$,

$$\frac{\tan(\theta) + \sin(\theta)}{\tan(\theta)} = \frac{2}{3}. \tag{9.34}$$

Solution:

We first consider the left hand side,

$$\frac{\tan(\theta) + \sin(\theta)}{\tan(\theta)} = \frac{\frac{\sin(\theta)}{\cos(\theta)} + \sin(\theta)}{\frac{\sin(\theta)}{\cos(\theta)}},$$

$$= \frac{\frac{\sin(\theta) + \cos(\theta)\sin(\theta)}{\cos(\theta)}}{\frac{\sin(\theta)}{\cos(\theta)}},$$

$$= \frac{\cos(\theta)\,(\sin(\theta) + \cos(\theta)\sin(\theta))}{\cos(\theta)\sin(\theta)},$$

$$= \frac{\sin(\theta) + \cos(\theta)\sin(\theta)}{\sin(\theta)},$$

$$= \frac{\sin(\theta)\,(1 + \cos(\theta))}{\sin(\theta)},$$

$$= 1 + \cos(\theta).$$

Using this result,

$$\frac{\tan(\theta) + \sin(\theta)}{\tan(\theta)} = \frac{2}{3},$$

$$\Rightarrow \qquad 1 + \cos(\theta) = \frac{2}{3}.$$

Solving (9.34) is therefore equivalent to solving,

$$\cos(\theta) = -\frac{1}{3}$$

Using a calculator we find that $\theta \approx 109.47°$. By the periodicity of the $\cos(\theta)$ function there is a further solution at $270 - (109.47 - 90) = 250.53°$.
Therefore, the full solution is given by $\theta \in \{109.47, 250.53\}$.

Example 9.8
Given that $x = 4\sin(\theta)$ and $y = 5\cos(\theta)$ show that $25x^2 + 16y^2 = 400$.

Solution:
We first rearrange,

$$x = 4\sin(\theta),$$

$$\Rightarrow \quad \frac{x}{4} = \sin(\theta),$$

$$y = 5\cos(\theta),$$

$$\Rightarrow \quad \frac{y}{5} = \cos(\theta).$$

Using Equation (9.2), we have

$$\sin^2(\theta) + \cos^2(\theta) = 1,$$

$$\Rightarrow \quad \left(\frac{x}{4}\right)^2 + \left(\frac{y}{5}\right)^2 = 1,$$

$$\Rightarrow \qquad \frac{x^2}{16} + \frac{y^2}{25} = 1,$$

$$\Rightarrow \qquad 25x^2 + 16y^2 = 400.$$

Example 9.9
Show that, for small angles, the function $f(x) = \sin^2(x)\cos(x)$ can be approximated by a function of the form $h(x) = A + Bx + Cx^2 + Dx^3 + Ex^4$ and use this approximation to evaluate $\sin^2\left(\frac{\pi}{24}\right)\cos\left(\frac{\pi}{24}\right)$.

Solution:

$$\sin^2(x)\cos(x) \approx (x)^2 \left(1 - \frac{x^2}{2}\right)$$

$$= x^2 - \frac{x^4}{2},$$

which is of the desired form with $A = B = D = 0$, $C = 1$ and $E = \frac{1}{2}$.
Using this approximation,

$$\sin^2\left(\frac{\pi}{24}\right)\cos\left(\frac{\pi}{24}\right) \approx \left(\frac{\pi}{24}\right)^2 - \frac{\left(\frac{\pi}{24}\right)^4}{2}$$

$$\approx 0.01699.$$

Note, the value obtained from a calculator is 0.0168913 to 7d.p.

Exercise 9.1

Q1. Express $\frac{\sin^2(\theta)}{\tan^2(\theta)}$ in terms of powers of $\sin(\theta)$ only.

Q2. Simplify $f(x) = \tan^2(x)\cos^4(x)$.

Q3. Solve $\cos(\theta)\tan(\theta) = \frac{\sqrt{3}}{2}$, giving all solutions in the range $0° \le \theta \le 360°$.

Q4. For $0° \le \theta \le 360°$ solve the following equation,

$$\frac{1 - 2\cos^2(\theta) + \cos^4(\theta)}{\sin^2(\theta)} = \frac{1}{2}.$$

Q5. (a) Prove

$$\frac{\cos^4(\theta) - \sin^4(\theta)}{\cos^2(\theta)} \equiv 1 - \tan^2(\theta)$$

(b) Hence, or otherwise, solve,

$$\frac{\cos^4(\theta) - \sin^4(\theta)}{\cos^2(\theta)} = \frac{1}{2},$$

giving all solutions in the range $0° \le \theta \le 540°$.

Q6. Express, solely in terms of $\cos(x)$,

$$f(x) = (\cos(x) + \sin(x))^3 - \sin(x)\left(2\cos^2(x) + 1\right)$$

Q7. Prove the following identities.

(a) $\dfrac{\sin(x)}{\cos^2(x)} = \sec(x)\tan(x)$

(b) $\tan(x) + \cot(x) = \sec(x)\csc(x)$

(c) $\csc(x) + \tan(x)\sec(x) = \csc(x)\sec^2(x)$

(d) $\sec^4(x) - \tan^4(x) = \sec^2(x) + \tan^2(x) = 1 + 2\tan^2(x)$

(e) $\sec^4(x) - \csc^4(x) = \dfrac{\sin^2(x) - \cos^2(x)}{\cos^4(x)\sin^4(x)}$

(f) $\dfrac{1}{\operatorname{cosec}(x) + \cot(x)} = \dfrac{1 - \cos(x)}{\sin(x)}$

(g) $\dfrac{\tan(x) + \cot(x)}{\sec(x) + \operatorname{cosec}(x)} = \dfrac{1}{\sin(x) + \cos(x)}$

Q8. (a) Using the diagram below provide an alternative proof for the identity,

$$\sin(x + y) = \sin(x)\cos(y) + \cos(x)\sin(y),$$

where x and y are both acute angles.

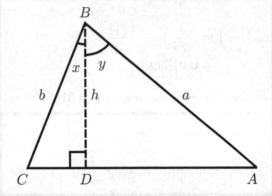

(b) Replacing y by $-y$ in your identity proved in part (a) find the corresponding identity for $\sin(x - y)$.

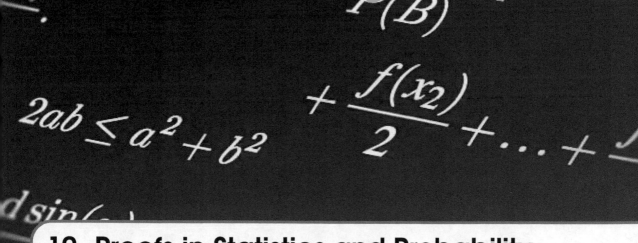

10. Proofs in Statistics and Probability

Statistics and Probability are areas of mathematics where special formulae are commonly used. Understanding the proofs of these formulae often involves manipulation of sums and series and provides useful practice of this skill.

.1 Descriptive Statistics

Suppose that we have a population and we would like to find some descriptive statistics. We recall the familiar definitions of mean, median and variance.

Definition 10.1 — Mean, Median and Variance

Let $\{x_1, x_2, \ldots, x_n\}$ be a population, ordered so that $x_1 \leq x_2 \leq \ldots \leq x_n$. The mean μ of the population is defined to be

$$\mu = \frac{1}{n} \sum_{i=1}^{n} x_i.$$

The median of the population is defined to be

$$\text{median} = \begin{cases} x_{\frac{n+1}{2}}, & n \text{ is odd}, \\ \frac{1}{2}\left(x_{\frac{n}{2}} + x_{\frac{n}{2}+1}\right), & n \text{ is even}. \end{cases}$$

The variance σ^2 is defined to be

$$\sigma^2 = \frac{1}{n} \sum_{i=1}^{n} (x_i - \mu)^2.$$

Formula 10.1 — Alternative Formula for Variance

An alternative formula for the variance of a sample is

$$\sigma^2 = \frac{1}{n} \sum_{i=1}^{n} x_i^2 - \mu^2.$$

Proof:

The proof follows by direct manipulation of the original formula for σ^2, and recalling the definition of μ:

$$\sigma^2 = \frac{1}{n} \sum_{i=1}^{n} (x_i - \mu)^2$$

$$= \frac{1}{n} \sum_{i=1}^{n} \left(x_i^2 - 2x_i \mu + \mu^2 \right)$$

$$= \frac{1}{n} \sum_{i=1}^{n} x_i^2 - 2\mu \left(\frac{1}{n} \sum_{i=1}^{n} x_i \right) + \frac{1}{n} \sum_{i=1}^{n} \mu^2$$

$$= \frac{1}{n} \sum_{i=1}^{n} x_i^2 - 2\mu^2 + \mu^2$$

$$= \frac{1}{n} \sum_{i=1}^{n} x_i^2 - \mu^2.$$

The mean and median can be characterised in alternatives ways, as the values which minimise the distance to all points in the population, for different definitions of distance. The following theorem and exercise show this.

Theorem 10.2 — Alternative Characterisation of the Mean

Let $\{x_1, x_2, \ldots, x_n\}$ be a population. Consider the function $f(x) : \mathbb{R} \mapsto \mathbb{R}$ defined by

$$f(x) = \sum_{i=1}^{n} (x_i - x)^2.$$

Then $f(x)$ has a unique minimum and the value of this minimum is μ, the mean. Hence, the mean is the point which minimises the sum of the square distances from each data point.

Proof:

We notice that $f(x)$ is a smooth function of x with x unbounded. We can therefore seek minima of f by finding x such that $\frac{df}{dx} = 0$. Now

$$\frac{df}{dx} = \frac{d\left(\sum_{i=1}^{n}(x_i - x)^2\right)}{dx}$$

$$\text{(linearity of differentiation)} = \sum_{i=1}^{n} \frac{d((x_i - x)^2)}{dx}$$

$$= \sum_{i=1}^{n} -2(x_i - x)$$

$$= 2\left(nx - \sum_{i=1}^{n} x_i\right).$$

Setting this equal to 0 gives

$$\frac{df}{dx} = 2\left(nx - \sum_{i=1}^{n} x_i\right) = 0.$$

$$\Rightarrow \qquad x = \frac{1}{n}\sum_{i=1}^{n} x_i$$

$$= \mu.$$

Hence, there is only one stationary point at $x = \mu$. Now $f(x) \to \infty$ as $x \to \pm\infty$, so it follows immediately that μ is the unique minimum of $f(x)$.

Proof Tip

In the previous proof, we could have differentiated $f(x)$ twice and shown that $\frac{d^2 f}{dx^2} > 0$ at $x = \mu$ to show that μ is a minimum. However, this would have been extra work and was unnecessary once the characteristics of f were taken into account.

Exercise 10.1

Suppose we have an ordered population $x_1, \ldots, x_n$ with

$$x_1 < x_2 < \cdots < x_n.$$

Let us define the function $g : \mathbb{R} \mapsto \mathbb{R}$ by

$$g(x) = \sum_{i=1}^{n} |x_i - x|.$$

(a) By considering the definition of $|\cdot|$, show that, for $j = 1, \ldots, n$,

$$g(x) = (2(j-1) - n)x + \sum_{i=j}^{n} x_i - \sum_{i=1}^{j-1} x_i.$$

(b) Now suppose that $n > 1$ is odd, show that for $x_{j-1} \leq x \leq x_j$, with $j = 2, \ldots (n+1)/2$, $g(x)$ is a linear function with negative gradient.

   Similarly, show that for $x_j \leq x \leq x_{j+1}$, with $j = (n+1)/2, \ldots, n-1$, $g(x)$ is a linear function with positive gradient.

   Hence, show that the median of the dataset minimises $g(x)$.

(c) Now suppose $n \geq 2$ is even, by working in a similar manner to part (a), find the *values* of x which minimise $g(x)$. Confirm that the median of the dataset minimises $g(x)$ in this case too.

The following exercise investigates what happens to the mean and variance of a data set under linear transformations.

Exercise 10.2

Given the data set $\{x_1, x_2, \ldots, x_N\}$, let $y_i = (x_i - a)/b$ where $b > 0$. Let μ_x and σ_x be the mean and standard deviation of x_i, and μ_y and σ_y be the mean and standard deviation of y_i.
Prove the following identities.

$$\mu_x = b\mu_y + a \qquad \text{and} \qquad \sigma_x = b\sigma_y.$$

10.2 Random Variables

Definition 10.3 — Expectation and Variance of Random Variables

Consider a random variable $X : \Omega \mapsto \mathbb{R}$, where Ω is a discrete sample space with n elements. Let the probability $\mathrm{P}(X = x_i) = p_i$, where x_i are possible values in the range of X. The expectation of X denoted $\mathbb{E}(X)$ is

$$\mathbb{E}(X) = \sum_{i=1}^{n} x_i p_i.$$

The variance of X, denoted $\mathrm{Var}(x)$ is

$$\mathrm{Var}(X) = \mathbb{E}(X^2) - (\mathbb{E}(X))^2 = \sum_{i=1}^{n} x_i^2 p_i - \left(\sum_{i=1}^{n} x_i p_i\right)^2.$$

Theorem 10.4 — Properties of Expectation

Suppose that $X_1, X_2, \ldots, X_n$ are independent random variables, then we have the following properties:

- $\mathbb{E}(X_1 + X_2 + \ldots + X_n) = \mathbb{E}(X_1) + \mathbb{E}(X_2) + \ldots + \mathbb{E}(X_n)$.
- $\mathbb{E}(X_i \cdot X_j) = \mathbb{E}(X_i) \cdot \mathbb{E}(X_j)$, for $i \neq j$.

We omit the proof of the above theorem, but use the results in what is to follow.

Example 10.1

Prove that the mean and variance of the binomial distribution $B(n,p)$ is given by np and $np(1-p)$ respectively.

Solution:

We recall that if $X \sim B(n,p)$, then the probability distribution of X is given by the formula:

$$P(X = k) = \binom{n}{k} p^k (1-p)^{n-k},$$

for $k = 0, \ldots, n$. The mean and the variance can be calculated directly from the definition of the binomial distribution, assuming the relevant properties of $\binom{n}{k}$ are known. However, it is simpler to use properties of expectation and an induction argument.

Let $P(n)$ be the statement:

"Let $X_n \sim B(n,p)$, then $\mathbb{E}(X_n) = np$ and $\text{Var}(X_n) = np(1-p)$"

Step 1:

The case $P(1)$ can be calculated directly. Suppose $X_1 \sim B(1,p)$ then

$$\mathbb{E}(X_1) = 0 \cdot \binom{1}{0} p^0 (1-p)^1 + 1 \cdot \binom{1}{1} p^1 (1-p)^0$$
$$= 0 \cdot (1-p) + 1 \cdot p = p,$$
$$\text{Var}(X_1) = \left(0^2 \cdot \binom{1}{0} p^0 (1-p)^1 + 1^2 \cdot \binom{1}{1} p^1 (1-p)^0 \right) - p^2$$
$$= p - p^2 = p(1-p).$$

Step 2:

We now assume that $P(k)$ is true for some $k \geq 1$.

Step 3:

To show that $P(k+1)$ is true, we first recall that, by the definition of the binomial distribution, if $X_{k+1} \sim B(k+1,p)$, then $X_{k+1} = X_k + X_1$, where $X_k \sim B(k,p)$ and $X_1 \ B(1,p)$ are independent. Hence, using the properties of expectation and the truth of $P(1)$ and $P(k)$, we have

$$\mathbb{E}(X_{k+1}) = \mathbb{E}(X_k + X_1)$$
$$= \mathbb{E}(X_k) + \mathbb{E}(X_1)$$
$$= kp + p = (k+1)p.$$

Similarly, for the variance we have

$$
\begin{aligned}
\mathrm{Var}(X_{k+1}) &= \mathrm{Var}(X_k + X_1) \\
&= \mathbb{E}\left((X_k + X_1)^2\right) - (\mathbb{E}(X_k + X_1))^2 \\
&= \mathbb{E}\left((X_k + X_1)^2\right) - (\mathbb{E}(X_k) + \mathbb{E}(X_1))^2 \\
&= \mathbb{E}\left(X_k^2 + 2X_1 X_k + X_1^2\right) - (\mathbb{E}(X_k))^2 - 2\mathbb{E}(X_k)\mathbb{E}(X_1) - (\mathbb{E}(X_1))^2 \\
&= \mathbb{E}\left(X_k^2\right) - (\mathbb{E}(X_k))^2 + \mathbb{E}\left(X_1^2\right) - (\mathbb{E}(X_1))^2 \\
&= \mathrm{Var}(X_k) + \mathrm{Var}(X_1) \\
&= kp(1-p) + p(1-p) = (k+1)p(1-p).
\end{aligned}
$$

Hence $P(k+1)$ is true.

Step 4:

By the principle of mathematical induction, we have that $P(n)$ is true for $n \geq 1$.

Suppose we have a population with some unknown mean μ and variance σ^2, where the population has a known distribution X. We would like to estimate the mean and variance and take a sample from X of size n with observations $x_1, x_2, \ldots, x_n$, then use these values to estimate the mean and the variance. In this case each x_i is drawn from the distribution $X_i \sim X$, with the X_i independent.

For example, we might wish to estimate the average height of females living in the United Kingdom, where the underlying distribution is assumed to be normal. It would be impossible to measure everybody, but by taking a sample we are able to estimate μ and σ^2.

Definition 10.5 — Sample Mean and Variance

For independent, identically distributed random variables $X_1, X_2, \ldots, X_n$, the sample mean $\bar{X}$ is

$$
\bar{X} = \frac{1}{n}\sum_{i=1}^{n} X_n.
$$

The sample variance S^2 is

$$
S^2 = \frac{1}{n-1}\sum_{i=1}^{n}(X_n - \bar{X})^2.
$$

Remark

From Definition 10.5, we see that that the sample mean and sample variance are both random variables. In order to find *estimates* for the true mean and variance, we use the same formulae, but with X_i replaced with the observed x_i. The sample mean and sample variance are *estimators*.

Remark
We note that the formula for the sample variance has a multiple of $\frac{1}{n-1}$, which is perhaps counterintuitive. We show why this is the case in the remaining work.

Definition 10.6 — Unbiased Estimator
Let $\hat{\theta}$ be an estimator of a parameter θ of some distribution. $\hat{\theta}$ is unbiased if $\mathbb{E}(\hat{\theta}) = \theta$.

The following example and exercise show that the sample mean and sample variance from Definition 10.5 are both unbiased estimators.

Example 10.2
Prove that the sample mean is an unbiased estimator for μ.

Solution:
We make use of the properties of expectation to show that this result is true. Recall that each $X_i \sim X$, then

$$
\begin{aligned}
\mathbb{E}\left(\frac{1}{n}\sum_{i=1}^{n} X_i\right) &= \frac{1}{n}\sum_{i=1}^{n} \mathbb{E}(X_i) \\
&= \frac{1}{n}\sum_{i=1}^{n} \mathbb{E}(X) \\
&= \frac{1}{n}\sum_{i=1}^{n} \mu \\
&= \frac{1}{n}\cdot n\mu \\
&= \mu.
\end{aligned}
$$

Hence, the sample mean is unbiased.

Exercise 10.3
In this exercise, we show that the sample variance is an unbiased estimator for the true variance of a random variable. This is a quite technical proof and requires the use of properties of expectation and summation notation.
(a) Show that

$$
\mathbb{E}(S^2) = \frac{1}{n-1}\sum_{i=1}^{n} \mathbb{E}\left((X_i - \bar{X})^2\right).
$$

(b) Hence, use the definition of $\bar{X}$ to show

$$
\mathbb{E}(S^2) = \frac{1}{n-1}\sum_{i=1}^{n} \mathbb{E}\left(X_i^2 - \frac{2}{n}X_i\sum_{j=1}^{n} X_j + \frac{1}{n^2}\left(\sum_{j=1}^{n} X_j\right)^2\right).
$$

(c) Show that

$$\frac{2}{n}X_i \sum_{j=1}^{n} X_j = \frac{2}{n}\left(X_i^2 + \sum_{\substack{j\neq i \\ j=1}}^{n} X_i X_j \right).$$

Hence, show that

$$\mathbb{E}\left(\frac{2}{n}X_i \sum_{j=1}^{n} X_j \right) = \frac{2}{n}\left(\mathbb{E}(X_i^2) + \sum_{\substack{j\neq i \\ j=1}}^{n} \mathbb{E}(X_i)\mathbb{E}(X_j) \right).$$

Finally, use $X_i \sim X$ to show

$$\mathbb{E}\left(\frac{2}{n}X_i \sum_{j=1}^{n} X_j \right) = \frac{2}{n}\mathbb{E}(X^2) + \frac{2(n-1)}{n}\left(\mathbb{E}(X) \right)^2.$$

(d) Show that

$$\frac{1}{n^2}\left(\sum_{j=1}^{n} X_j \right)^2 = \frac{1}{n^2}\left(\sum_{l=1}^{n}(X_l)^2 + \sum_{l=1}^{n}\sum_{\substack{j\neq l \\ j=1}}^{n} X_l X_j \right).$$

Hence, use properties of expectation to show

$$\mathbb{E}\left(\frac{1}{n^2}\left(\sum_{j=1}^{n} X_j \right)^2 \right) = \frac{1}{n^2}\left(\sum_{l=1}^{n}\mathbb{E}(X_l^2) + \sum_{l=1}^{n}\sum_{\substack{j\neq l \\ j=1}}^{n} \mathbb{E}(X_l)\mathbb{E}(X_j) \right).$$

Finally, use $X_i \sim X$ to show

$$\mathbb{E}\left(\frac{1}{n^2}\left(\sum_{j=1}^{n} X_j \right)^2 \right) = \frac{1}{n}\mathbb{E}(X^2) + \frac{n-1}{n}\left(\mathbb{E}(X) \right)^2.$$

(e) Combine the results of parts (c) and (d) to show

$$\mathbb{E}\left((X_i - \bar{X})^2 \right) = \frac{n-1}{n}\left(\mathbb{E}(X^2) - (\mathbb{E}(X))^2 \right).$$

Hence, conclude that

$$\mathbb{E}(S^2) = \mathrm{Var}(X)$$

and S^2 is an unbiased estimator for the variance.

Proof Tip

Often it is useful to use our intuition when developing a proof. Sadly, the proofs of the unbiased nature of the sample mean and sample variance do not provide much intuition as to why they are unbiased. Consider the figure below, which shows the same population, with mean μ and range R, and two possible samples. The horizontal black line represents the values of some population. The red boxes indicate the range of values which are contained in each sample.

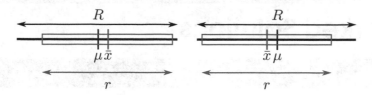

We notice that the mean of the sample $\bar{x}$ can be larger or smaller than the actual mean μ, half of the time it will be bigger, the other half smaller, on average, it will have the value μ.

If r is the true range of the sample data, we see that the maximum value it can ever take is R. However, most of the time, it will be smaller than R. This means that, on average, the value of r is less than R and it is biased. To make the r unbiased, we could multiply it by some factor > 1. As the variance is an alternative measure of the spread of the data, it follows that we must also multiply the true variance of the sample data by a factor > 1. This factor turns out to be $\frac{n}{n-1} > 1$.

Worked Solutions — Exercise 3.1

Q1. (a) This is a statement that is false because some dogs do not have tails;

(b) This is a statement that is true, there certainly is a donkey somewhere;

(c) This is neither a statement, nor a predicate: a monkey cannot be true or false;

(d) This is neither a statement nor a predicate: a variable cannot be true or false;

(e) This is neither a statement nor a statement;

(f) This is a predicate, because it can be true or false depending on the value of θ;

(g) This is a true statement;

(h) This is a false statement;

(i) This is a predicate, because it can be true or false depending on the value of x;

(j) This is a predicate, because we do not know what p and q are;

(k) This is a statement which is true;

(l) This is a statement which is false.

Q2. (a) There is a car with no handbrake;

(b) There is a car that does not have a handbrake that is guaranteed to stop the car;

(c) All flights have at least one seat where smoking is not allowed;

(d) There is a town where in all the restaurants there is an item on the menu which does not contain potatoes;

(e) There is somebody who does not buy *Stoves Today* who neither reads it, nor uses it for kindling.

Q3. (a) $\exists x\, p(x)$;

(b) $\forall x\, p(x)$;

(c) $\forall x\, \exists y\, \neg p(x, y)$;

(d) $\exists x\, \forall y\, \forall z\, \neg p(x, y, z)$;

(e) $\forall y\, \exists x\, \exists z\, p(x, y, z)$;

(f) $\forall x\, \exists y\, (\neg p(x) \wedge \neg q(y))$;

(g) $\exists z\, \exists y\, \exists x\, (\neg p(x) \vee (\neg q(y) \wedge r(z)))$;

(h) $(\forall x\, \neg p(x)) \vee (\forall y\, p(y))$.

Worked Solutions — Exercise 3.2

The truth table is shown below

A	B	$\neg A$	$\neg B$	$A \Rightarrow B$	$\neg B \Rightarrow \neg A$
T	T	F	F	T	T
T	F	F	T	F	F
F	T	T	F	T	T
F	F	T	T	T	T

We conclude that, because for all possibilities, the truth values for $A \Rightarrow B$ and $\neg B \Rightarrow \neg A$ are the same, that a contrapositive of a statement is equivalent to the statement.

Worked Solutions — Exercise 3.3

Q1. (a) Every square is a rhombus, but not every rhombus is a square. Hence,

$$S \text{ is a rhombus} \Leftarrow S \text{ is a square.}$$

(b) Every rhombus is a parallelogram, but not every parallelogram is a rhombus. Hence

$$X \text{ is a rhombus} \Rightarrow X \text{ is a parallelogram.}$$

(c) Since n is an integer, the compound statement "$n^2 > 8$ and $n > 0$" is equivalent to $n \geq 3$, since 3 is the least integer greater than or equal to $\sqrt{8}$. Hence,

$$n^2 > 8 \text{ and } n > 0 \Leftrightarrow n \geq 3.$$

(d) It is clear that any rational number squared is rational, since for integer p, q with $q \neq 0$,

$$\left(\frac{p}{q}\right)^2 = \frac{p^2}{q^2}.$$

However, it is false that every rational number has rational square roots. For example,

$$\left(\sqrt{2}\right)^2 = 2,$$

which is rational, but $\sqrt{2}$ is irrational. Hence,

$$x^2 \text{ is rational} \Leftarrow x \text{ is rational.}$$

Q2. (a) This is a valid argument. P1 and P2 can be written as

$$\text{P1:}\quad \text{Cats} \Rightarrow \text{Wings}, \qquad \text{P2:}\quad \text{Wings} \Rightarrow \text{Four legs}.$$

Combining P1 and P2, we obtain Q. Regardless of whether P1 and P2 are true, the conclusion Q follows from P1 and P2, so the argument is valid.

(b) This is not a valid argument. The statement P1 is true for all real x and y. The inequality $x^2 + y^2 \leq 1$ describes the interior and boundary of a circle centred at the origin with radius 1. Hence, $-1 \leq x \leq 1$ in this region. Statement P2 forces $x = 1/4$.

Now, let $x = 1/4$, $y = 2$. P1 and P2 are satisfied (since P1 is a one-way implication), but Q is not satisfied. Since there is at least one counter example, the argument is invalid.

(c) This is a valid argument. P1 and P2 can be written as

$$\text{P1:}\quad y \leq 4 \Rightarrow x \geq 3, \qquad \text{P2:}\quad y \leq 4 \Rightarrow z^2 > x.$$

Combining P1 and P2, we obtain $y \leq 4 \Rightarrow z^2 > x \geq 3$ and so

$$y \leq 4 \Rightarrow z^2 > 3.$$

We can read this statement as "$y \leq 4$ only if $z^2 > 3$". In particular, if $z = 0 \leq 3$, then we must have $y > 4$, *i.e.* Q follows from P1 and P2.

(d) This is not a valid argument. For a counter example, take $x = 1$, $y = 6$, $z = 2$. Then P1 and P2 are satisfied, but Q is not, so Q does not follow from P1 and P2.

(e) This is not a valid argument. For a counter example, take $x = 5$, $y = 4$, $z = 6$. P1 and P2 are satisfied, but Q is not, so Q does not follow from P1 and P2.

Q3. Following the hint, we use the following labels: B (eats bagels), S (can swim), C (has a calculator), P (plays poker), A (has an abacus), F (likes fencing).

(a) "No Mathtopian that eats bagels cannot swim."

This statement says that there is no such thing as a Mathtopian that eats bagels and cannot swim. Hence, if a Mathtopian eats bagels, then they can swim:

$$B \Rightarrow S$$

(b) "No Mathtopian without a calculator plays poker."

This statement says that there is no such thing as a Mathtopian that does not own a calculator and plays poker. That is, if a Mathtopian does not own a calculator, then they do not play poker. This can be written as $\neg C \Rightarrow \neg P$. An equivalent statement can be produced by negating both sides, then reversing the implication. This is called taking the *contrapositive*. Then we obtain

$$P \Rightarrow C.$$

(c) "Mathtopians who have an abacus all eat bagels."

This statement is simply

$$A \Rightarrow B.$$

(d) "No Mathtopian who can swim likes fencing."

This statement says that there is no such thing as a Mathtopian that can swim that likes fencing. Equivalently, if a Mathtopian can swim, they do not like fencing. We write this as

$$S \Rightarrow \neg F.$$

(e) "No Mathtopian has a calculator unless they have an abacus."

Using the hint, we write "A unless B" as $\neg B \Rightarrow A$. The statement says there is no such thing as a Mathtopian that has a calculator unless they have an abacus, *i.e.* "$\neg C$ unless A", or $\neg A \Rightarrow \neg C$. Taking the contrapositive (see step 2), we obtain

$$C \Rightarrow A.$$

We can now write a chain of implications using statements 1-5 to obtain

$$P \Rightarrow C \Rightarrow A \Rightarrow B \Rightarrow S \Rightarrow \neg F.$$

Thus, we can legitimately conclude that $P \Rightarrow \neg F$, *i.e.* a Mathtopian that plays poker does not like fencing.

Worked Solutions — Exercise 4.1

Q1. Expanding the left-hand side, we obtain

$$(x + y)(x - y) = x^2 - xy + xy - y^2 = x^2 - y^2.$$

Hence, we have shown directly that the left-hand side is equivalent to the right-hand side.

Q2. If $(x - p)$ is a factor of the polynomial expression $f(x)$, then there exists a polynomial expression $g(x)$ such that $f(x) = (x - p)g(x)$. Then

$$f(p) = (p - p)g(p) = (0)g(p) = 0.$$

Q3. Factorising,

$$x^3 + x^2 - 2x - 8 = (x - 2)(x^2 + 3x + 4).$$

The quadratic equation $x^2 + 3x + 4 = 0$ has discriminant $b^2 - 4ac = 3^2 - 4(1)(4) = -7 < 0$, and so has no real roots. Hence, the expression $x^3 + x^2 - 2x - 8$ has only one real factor.

Q4. The quadratic $kx^2 + 2kx - 3 = 0$ has discriminant $b^2 - 4ac = 4k^2 - 4(k)(-3) = 4k(k + 3)$. If the quadratic has no real roots, then k satisfies

$$4k(k + 3) < 0 \Rightarrow -3 < k < 0.$$

Q5. The quadratic $kx^2 - 3x + k = 0$ has discriminant $b^2 - 4ac = (-3)^2 - 4(k)(k) = 9 - 4k^2$. If the quadratic has two distinct real roots, then k satisfies

$$9 - 4k^2 > 0 \Rightarrow -\frac{3}{2} < k < \frac{3}{2}.$$

Q6. (a) Consider the diagram below.

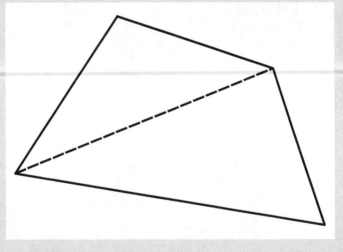

Any convex quadrilateral can be divided into two triangles. Since the interior angles of a triangle sum to $180°$, the interior angles of a convex quadrilateral sum to $2 \times 180° = 360°$.

(b) Extending the proof in part **(a)**, any n-sided convex polygon can be divided into $n - 2$ triangles. Hence, the interior angles of an n-sided polygon sum to $180(n - 2)°$.

Q7. Completing the square,

$$x^2 - 6x + 10 = (x - 3)^2 + 1.$$

Since $(x - 3)^2 \geq 0$ for all real x, $(x - 3)^2 + 1 \geq 1$ for all real x.

Q8. The coordinates of a point where the line and circle intersect satisfies $y = mx + 2$ and $(x + 1)^2 + (y - 2)^2 = r$ simultaneously. Substituting $y = mx + 2$ into the circle equation,

$$(x + 1)^2 + (mx)^2 = r,$$
$$\Rightarrow x^2 + 2x + 1 + m^2x^2 - r = 0,$$
$$\Rightarrow (1 + m^2)x^2 + 2x + 1 - r = 0.$$

If the line and circle intersect at two distinct points, this quadratic has two distinct real roots. Hence,

$$b^2 - 4ac = 4 - 4(1 + m^2)(1 - r) > 0,$$
$$\Rightarrow 4 - 4 - 4m^2 + 4r + 4rm^2 > 0,$$
$$\Rightarrow r(1 + m^2) > m^2,$$
$$\Rightarrow r > \frac{m^2}{1 + m^2}.$$

Q9. The function $f(x) = 3x - 1$ crosses the x-axis when $3x - 1 = 0$. This is a linear equation in x, with the unique solution $x = 1/3$. Hence, $f(x)$ intersects the x-axis exactly once, at $x = 1/3$.

We can also use the derivative of f to prove the uniqueness of the x-axis crossing. We have $f'(x) = 3$. Since the gradient of f is constant and positive for all real x, the function is always increasing and so has no turning points. Hence, f only crosses the x-axis once.

Q10. The function $f(x) = x^2 - 5$ crosses the x-axis when $x^2 - 5 = 0$. This is a quadratic equation in x, with the real distinct solutions $x = \pm\sqrt{5}$. Hence, $f(x)$ intersects the x-axis exactly twice, at $x = \pm\sqrt{5}$.

Similar to the previous question, we can check that there are no more x-axis crossings using $f'(x) = 2x$. The equation $f'(x) = 0$ has the unique solution $x = 0$, so $f(x)$ has only one turning point (at $x = 0$). Since there are no more turning points, there can be no further x-axis crossings.

Q11. The function $f(x) = x^3 - x - 1$ is a cubic, so potentially has 3 real roots. Since $f(1) = -1$ and $f(2) = 5 > 0$, f has a root between $x = 1$ and $x = 2$. We must now show that there are no further real roots.

Consider $f'(x) = 3x^2 - 1$. Now, $3x^2 - 1 = 0$ has solutions $x = \pm\dfrac{1}{\sqrt{3}}$. Since

$$f\left(\frac{1}{\sqrt{3}}\right) = \frac{-4 - 3\sqrt{3}}{3\sqrt{3}} < 0, \quad f\left(-\frac{1}{\sqrt{3}}\right) = \frac{2 - 3\sqrt{3}}{3\sqrt{3}} < 0,$$

both stationary points are located beneath the x-axis. We also have $f''(x) = 6x$, so $f''(1/\sqrt{3}) = 6/\sqrt{3} > 0$ and $f''(-1/\sqrt{3}) = -6/\sqrt{3} < 0$. Hence, f has a local maximum at $x = -1/\sqrt{3}$ and a local minimum at $x = 1/\sqrt{3}$. Since both stationary points are beneath the x-axis and neither are points of inflection, f can only cross the x-axis once, so there are no more real roots other than the one between $x = 1$ and $x = 2$.

Q12. Multiplying the numerator and denominator of $f(x)$ by $\sqrt{a+x} + \sqrt{a}$,

$$
\begin{aligned}
f(x) &= \frac{(\sqrt{a+x} - \sqrt{a})(\sqrt{a+x} + \sqrt{a})}{x(\sqrt{a+x} + \sqrt{a})}, \\
&= \frac{a + x - a}{x(\sqrt{a+x} + \sqrt{a})}, \\
&= \frac{x}{x(\sqrt{a+x} + \sqrt{a})}, \\
&= \frac{1}{\sqrt{a+x} + \sqrt{a}}.
\end{aligned}
$$

Hence,

$$
\lim_{x \to 0} f(x) = \frac{1}{\sqrt{a+0} + \sqrt{a}} = \frac{1}{2\sqrt{a}}.
$$

Q13. Rearranging the expression for a_n, we obtain

$$
a_n = 2 - \frac{\sqrt{n}}{n} = 2 - \frac{1}{\sqrt{n}}.
$$

Since $n \in \mathbb{N}$, we have $n > 1 > 0$ and so $\dfrac{1}{\sqrt{n}} > 0$ and $a_n < 2$ for all n.

Q14. Using the hint, consider the number $\sqrt{2}^{\sqrt{2}}$. It is unclear whether this is rational or irrational. However, it does not matter! If $\sqrt{2}^{\sqrt{2}}$ is rational, the proof is complete, with $a = b = \sqrt{2}$ (irrational) and a^b rational. If $\sqrt{2}^{\sqrt{2}}$ is irrational, we set $a = \sqrt{2}^{\sqrt{2}}$, $b = \sqrt{2}$. Then

$$
a^b = \left(\sqrt{2}^{\sqrt{2}} \right)^{\sqrt{2}} = \left(\sqrt{2} \right)^2 = 2.
$$

Clearly, 2 is rational, so we have found irrational a and b such that a^b is rational and thus the proof is complete.

Q15. (a) This inequality is true. We are given that $a < b$. Adding c to both sides of the inequality,

$$
a + c < b + c.
$$

Since $c < d$, $b + c < b + d$, and so

$$
a + c < b + d.
$$

(b) This inequality is true. We are given that $a < b$, so subtracting c from both sides yields

$$a - c < b - c.$$

(c) This inequality is false. As a counter-example, take $a = -2$, $b = 1$, $c = -2$, $d = 1$, so $a < b$, $c < d$ are both satisfied. However,

$$ac = 4 \not< 1 = bd.$$

(d) This inequality is false. Take $a = 4$, $b = 6$, $c = 1$, $d = 3$, so $a < b$, $c < d$ are both satisfied. However,

$$\frac{a}{c} = 4 \not< 2 = \frac{b}{d}.$$

(e) This inequality is false. Take $a = 2$, $b = 3$, $c = -1$, $d = 1$, so $a < b$, $c < d$ are both satisfied. However,

$$a - \frac{1}{c} = 3 \not< 2 = b - \frac{1}{d}.$$

(f) This inequality is true. The graph of x^3 is strictly increasing (this can be seen clearly by observing the graph of $y = x^3$), so $a < b \Rightarrow a^3 < b^3$.
Alternatively, we note that $a^3 < b^3$ is true if $b^3 - a^3 > 0$, so $(b - a)(b^2 + ba + a^2) > 0$. The last inequality is true, since $a < b \Rightarrow b - a > 0$ and $b^2 + ba + a^2 = (b + \frac{a}{2})^2 + \frac{3}{4}a^2 > 0$.

(g) This inequality is true. In part (f), we showed that $a < b \Rightarrow a^3 < b^3$, $c < d \Rightarrow c^3 < d^3$. The inequality in (a) then gives $a^3 + c^3 < b^3 + d^3$. Rearranging,

$$a^3 - d^3 < b^3 - c^3.$$

Q16. We first consider the denominators of each side of the inequality.

$$x^2 + 2x + 3 = (x + 1)^2 + 2 > 0,$$
$$x^2 + 4x + 5 = (x + 2)^2 + 1 > 0,$$

for all $x \in \mathbb{R}$. Hence, we can cross-multiply the fractions without changing the inequality. This yields

$$(x - 1)(x^2 + 4x + 5) \leq (x + k)(x^2 + 2x + 3).$$

Expanding the brackets and simplifying,

$$x^3 + 3x^2 + x - 5 \leq x^3 + (k + 2)x^2 + (2k + 3)x + 3k,$$
$$\Rightarrow F(x) = (k - 1)x^2 + (2k + 2)x + 3k + 5 \geq 0.$$

Recall that for a general quadratic $ax^2 + bx + c$ to have constant sign for all real x, we require the discriminant to be less than or equal to zero, so $b^2 - 4ac \leq 0$.

In this case, the quadratic $F(x)$ has $a = k - 1$, $b = 2k + 2$, $c = 3k + 5$. Hence, we require

$$(2k + 2)^2 - 4(k - 1)(3k + 5) = -8k^2 + 24 \leq 0 \Rightarrow k^2 \geq 3.$$

This is satisfied for $k \leq -\sqrt{3}$ or $k \geq \sqrt{3}$. We disregard $k \leq -\sqrt{3}$, since $k - 1 < 0$ in this case and $F(x) < 0$ for large x, and we require $F(x) \geq 0$ for all real x. Thus, the set of values of k for which

$$\frac{x - 2}{x^2 + 2x + 3} \leq \frac{x + k}{x^2 + 4x + 5}$$

holds for all $x \in \mathbb{R}$ is $k \geq \sqrt{3}$.

Q17. Using basic trigonometry, we have $a = \sin\theta$, $b = \cos\theta$. Hence we can write $\theta = \arcsin a = \arccos b$. Furthermore, we can write $a = \sin(\arccos b)$, $b = \cos(\arcsin a)$. Using Pythagoras' theorem, we can also write $b = \sqrt{1 - a^2}$, $a = \sqrt{1 - b^2}$. Hence,

$$a = \sqrt{1 - b^2} = \sin(\arccos b), \quad b = \sqrt{1 - a^2} = \cos(\arcsin a).$$

Generalising,

$$\cos(\arcsin x) = \sin(\arccos x) = \sqrt{1 - x^2}.$$

Q18. (a) Following the suggestion in the question, we have

$$\sum_{k=1}^{n} (xa_k - b_k)^2 \geq 0.$$

Expanding,

$$\sum_{k=1}^{n} (xa_k - b_k)^2 = Ax^2 - 2Bx + C \geq 0,$$

where

$$A = \sum_{k=1}^{n} a_k^2, \quad B = \sum_{k=1}^{n} a_k b_k, \quad C = \sum_{k=1}^{n} b_k^2.$$

For the quadratic to have constant sign, we require $(-2B)^2 - 4AC \leq 0 \Rightarrow B^2 \leq AC$. Hence, we have deduced the Cauchy-Schwarz inequality:

$$\left(\sum_{k=1}^{n} a_k b_k \right)^2 \leq \left(\sum_{k=1}^{n} a_k^2 \right) \left(\sum_{k=1}^{n} b_k^2 \right).$$

(b) Rearranging the given inequality, we obtain

$$n^2 \leq (x_1 + \ldots + x_n) \left(\frac{1}{x_1} + \ldots + \frac{1}{n} \right).$$

Using the hint, we note that

$$n^2 = \left(\sum_{k=1}^{n} 1 \right)^2.$$

Letting $a_k = \sqrt{x_k}$ and $b_k = \dfrac{1}{\sqrt{x_k}}$ in the Cauchy-Schwarz inequality, we obtain

$$n^2 = \left(\sum_{k=1}^{n} 1 \right)^2 = \left(\sum_{k=1}^{n} \sqrt{x_k} \cdot \frac{1}{\sqrt{x_k}} \right)^2 \le (x_1 + \ldots + x_n) \left(\frac{1}{x_1} + \cdots \frac{1}{x_n} \right).$$

Q19. (a) We have

$$(\sqrt{x} - \sqrt{y})^2 = x - 2\sqrt{xy} + y \ge 0.$$

Hence,

$$\frac{x+y}{2} \ge \sqrt{xy},$$

so we have deduced the arithmetic-geometric mean (AM-GM) inequality for two numbers.

(b) Given that $a+b+c = 1$, we have $a - 1 = b + c$. Using the AM-GM inequality, $b + c \ge 2\sqrt{bc}$. Similarly, $b - 1 \ge 2\sqrt{ac}$, $c - 1 \ge 2\sqrt{ab}$. Hence

$$\left(\frac{1-a}{a} \right) \left(\frac{1-b}{b} \right) \left(\frac{1-c}{c} \right) \ge \frac{(2\sqrt{bc})(2\sqrt{ac})(2\sqrt{ab})}{abc},$$

$$= \frac{8\sqrt{a^2 b^2 c^2}}{abc},$$

$$= 8.$$

Hence,

$$\left(\frac{1-a}{a} \right) \left(\frac{1-b}{b} \right) \left(\frac{1-c}{c} \right) \ge 8,$$

as required.

Worked Solutions — Exercise 4.2

Q1. (a) Let $n = 7$. Then $7! = 5040 > 2187 = 3^7$, so the statement is false.

(b) Consider $30 = 2 \times 3 \times 5$, which has an odd number of distinct prime factors. Hence, the statement is false.

(c) Let $n = 1$. Then $n^2 + 2n - 2 = 1$, which is odd. Hence, the statement is false.

(d) Let $m = 0$, $c \ne 0$. The straight line $y = c$ is parallel to the x-axis and so does not intersect it. Hence, the statement is false.

Q2. (a) Since a and b are both positive integers, $a, b \geq 1$. Rearranging,

$$\frac{a}{b} + \frac{b}{a} = \frac{a^2 + b^2}{ab}.$$

Since $a, b \geq 1$, $a^2 + b^2 \geq 2$ and $ab \geq 1$. Thus,

$$\frac{a^2 + b^2}{ab} \geq 2.$$

(b) Let $a = 1$, $b = -2$. Then

$$\frac{a}{b} + \frac{b}{a} = -\frac{1}{2} - 2 = -\frac{5}{2} < 2,$$

so the statement is false when at least one of a or b is negative.

Q3. (a) If n is odd, then there exists a positive integer p such that $n = 2p - 1$. Then

$$n^2 = (2p - 1)^2 = 4p^2 - 2p + 1 = 2(p^2 - p) + 1.$$

Since $p^2 - p$ is a positive integer, n^2 has the form of an odd integer, so if n is odd, n^2 is odd.

(b) If m and n are odd, there exist positive integers p and q such that $m = 2p - 1$, $n = 2q - 1$. Then

$$mn = (2p - 1)(2q - 1) = 4pq - 2p - 2q + 1 = 2(2pq - p - q) + 1.$$

Since $2pq - p - q$ is a positive integer, mn has the form of an odd integer. Hence, if m and n are odd, mn is odd.

(c) If m and n are even, then there exist positive integers p and q such that $m = 2p$ and $n = 2q$. Then

$$mn = (2p)(2q) = 4pq = 2(2pq).$$

Since $2pq$ is a positive integer, mn has the form of an even number. Thus, if m and n are even, mn is even.

(d) If m is odd and n is even, then there exist positive integers p and q such that $m = 2p - 1$ and $n = 2q$. Then

$$mn = (2p - 1)2q = 4pq - 2q = 2(2pq - q).$$

Since $2pq - q$ is a positive integer, mn has the form of an even number. Hence, if m is odd and n is even, mn is even.

(e) If n is even, then there exists a positive integer p such that $n = 2p$. Then

$$7(n + 4) = 7(2p + 4) = 14p + 28 = 2(7p + 14).$$

Since $7p + 14$ is a positive integer, $7n + 4$ has the form of an even number. Hence, if n is even, $7n + 4$ is even.

Note that we could have used the previously proven facts that the product of an even and odd integer is even, and the sum of two even numbers is even to deduce this result.

Q4. (a) If n^2 is odd, then n is odd.

This is true. We found in Q1 that a product involving at least one even number is even. Hence, if n^2 is odd, n must be odd.

(b) If mn is odd, m and n are odd.

This is also true, with the same reasoning as in **(a)**.

(c) If mn is even, then m and n are even.

This is false. For a counter example, consider $m = 2$, $n = 3$. Then $mn = 6$ is even, but n is odd.

(d) If mn is even, then m is odd and n is even.

This is false. Consider $m = n = 2$. Then $mn = 4$ is even, but m is even.

(e) If $7n + 4$ is even, then n is even.

This is true. If $7n + 4$ is even, then there exists a positive integer q such that $7n + 4 = 2p$. Then

$$7n + 4 = 2p \Rightarrow 7n = 2p - 4 = 2(p - 2).$$

Since $2(p - 2)$ is even, we must have $7n$ even for this equation to make sense. Since 7 is clearly odd, we conclude that n is even.

Q5. Let a and b be positive integers with $a \neq b$. Then the square of the sum of a and b is

$$(a + b)^2 = a^2 + b^2 + 2ab > a^2 + b^2,$$

so the square of the sum of a and b is greater than the sum of their squares. If $a = b$, then

$$(2a)^2 = 4a^2 > a^2 + a^2 = 2a^2,$$

so the statement is also true for $a = b$.

Q6. (a) Let m and n be two consecutive even numbers. Then there exists positive integer k such that $m = 2k$ and $n = 2(k + 1)$. Then

$$n^2 - m^2 = 4(k + 1)^2 - 4k^2 = 4(2k + 1).$$

Hence, the statement is true.

(b) Let m and n be two consecutive odd numbers. Then there exists positive integer k such that $m = 2k - 1$, $n = 2k + 1$. Then

$$n^2 - m^2 = 4k^2 + 4k + 1 - (4k^2 - 4k + 1) = 4(2k).$$

Hence, the statement is true.

(c) Let $m = 2$, $n = 3$. Then

$$n^2 - m^2 = 9 - 4 = 5.$$

Since 5 is not a multiple of 4, the statement is false.

Q7. Let $n = 5$. Then $5! - 1 = 119 = 7 \times 17$. Hence, the statement is false.

Q8. (a) If b and c are divisible by a, then there exist positive integers p and q such that $b = ap$ and $c = aq$. Then

$$b + c = ap + aq = a(p + q),$$

so $b + c$ is also divisible by a.

(b) If b is divisible by a and c is divisible by b, then there exist positive integers p and q such that $b = ap$, $c = qb$. Then

$$c = qb = q(ap) = a(qp).$$

Since qp is an integer, c is divisible by a.

Q9. The first point to note is that $4a + 3$ is odd. Using previous results, $4a$ is even regardless of whether a is odd or even, then $4a + 3$ is odd (even + odd = odd). Hence, we need $b^2 + c^2$ to be odd for the equation to be valid. We have already proven in Q3 that if n is odd, then n^2 is odd. Similarly, if n is even, n^2 is even. Also, odd + odd = even, and even + even = even. Hence, we must have c odd and b even, say.

Thus, there exist positive integers p and q such that $c = 2p - 1$ and $b = 2q$. Then,

$$4a + 3 = b^2 + c^2 = (2p-1)^2 + (2q)^2 = 4p^2 - 4p + 1 + 4q^2 = 4(p^2 + q^2 - p) + 1.$$

The equation is inconsistent. Although both sides of the equations are odd, the left-hand side has remainder 3 when divided by 4. However, since $p^2 + q^2 - p$ is an integer, the right-hand side has remainder 1 when divided by 4. Hence, there are no integers a, b and c for which the equation holds.

Q10. (a) Following the hint, let the rectangle have integer side lengths x and y such that $x \leq y$. If the perimeter is equal to the area, then we require

$$2x + 2y = xy. \tag{11.1}$$

Dividing (11.1) by $2xy$ gives

$$\frac{1}{x} + \frac{1}{y} = \frac{1}{2}. \tag{11.2}$$

Since $x \leq y$, (11.2) is only satisfied for x at most 4. If x is 5 or greater, then so is y and the left-hand side of (11.2) is at most $2/5 < 1/2$.

The choice of $x = 4$ gives $y = 4$ as one solution. Next, $x = 3$ gives $y = 6$.

Finally, $x \leq 2$ is impossible, since $1/y$ is positive.

Hence, the only rectangles with integer side lengths that have the property that their perimeter is equal to their area are the 3×6 rectangle and the 4×4 square.

An alternative method is to rewrite (11.1) as

$$xy - 2x - 2y + 4 = 4,$$

which factorises to give

$$(x - 2)(y - 2) = 4. \tag{11.3}$$

Both factors on the left-hand side of (11.3) must be integers and factors of 4. Since $4 = 2 \times 2 = 4 \times 1$, the only solutions we get, with $x \le y$, are

$$x - 2 = 2 = y - 2 \implies x = y = 4,$$
$$x - 2 = 1, \ y - 2 = 4 \implies x = 3, \ y = 6,$$

as before.

(b) Following the hint again, let the cuboid have integer side lengths p, q, r such that $p \le q \le r$. If the total surface area is equal to twice the volume, we obtain

$$2pq + 2qr + 2rp = 2pqr. \tag{11.4}$$

Dividing (11.4) by $2pqr$ gives

$$\frac{1}{p} + \frac{1}{q} + \frac{1}{r} = 1. \tag{11.5}$$

Since $p \le q \le r$, (11.5) is only satisfied for $p \le 3$. If $p \ge 4$, then then so is q and r, so the left-hand side of (11.5) is at most $3/4 < 1$. If $p = 3$, then we must have $q = r = 3$, so that (11.5) becomes

$$\frac{1}{3} + \frac{1}{3} + \frac{1}{3} = 1.$$

If $p = 2$, then (11.5) is satisfied for

$$\frac{1}{q} + \frac{1}{r} = \frac{1}{2}.$$

This is equivalent to the 2D problem we have already solved above, which has solutions $q = r = 4$ and $q = 3$, $r = 6$.

Finally, $p = 1$ is impossible, since $1/q$ and $1/r$ are positive.

Thus, the only cuboids with integer side lengths that have the property that their surface area is twice their volume are the $3 \times 3 \times 3$ cube, and the $2 \times 4 \times 4$ and $2 \times 3 \times 6$ cuboids.

Worked Solutions — Exercise 5.1

Q1. For $x \in \mathbb{Z}$ the contrapositive of "if $x^2 - 8x + 7$ is even then x is odd" is the statement "If x is even then $x^2 - 8x + 7$ is odd".

To prove this, let $x = 2m$ for some integer m. Then,

$$
\begin{aligned}
x^2 - 8x + 7 &= (2m)^2 - 8(2m) + 7 \\
&= 4m^2 - 16m + 7 \\
&= 2(2m^2 - 8m + 3) + 1,
\end{aligned}
$$

which is odd.

Therefore, we have proven the contrapositive statement and so can now infer the truth of our original statement.

Q2. For $x, y \in \mathbb{Z}$ the contrapositive of the statement "xy is even, then at least one of x or y must be even" is "not at least one of x or y is even, then xy is odd". This contrapositive can be re-written in a more natural way as "Both x and y are odd, then xy is odd."

To prove the contrapositive let $x = 2m + 1$ and $y = 2n + 1$ for some integers m and n. Then,

$$
\begin{aligned}
xy &= (2m + 1)(2n + 1) \\
&= 4mn + 2m + 2n + 1 \\
&= 2(2mn + m + n) + 1,
\end{aligned}
$$

which is odd.

Therefore, we have proven the contrapositive statement and so can now infer the truth of our original statement.

Worked Solutions — Exercise 5.2

Q1. Suppose, for contradiction that there exist two even numbers whose sum is odd. Then, for positive integers a and b, and positive odd integer c, we have

$$
\begin{aligned}
& 2a + 2b = c, \\
\Rightarrow\quad & 2(a + b) = c, \\
\Rightarrow\quad & a + b = \frac{c}{2}.
\end{aligned}
$$

Since we assumed that c is odd, it is not divisible by 2, so $\frac{c}{2}$ is not an integer. However, $a + b$ *is* an integer, so we have arrived at a contradiction. Hence, there are no such even integers whose sum is odd and we conclude that the sum of any two even numbers is even.

Q2. Suppose, for contradiction, that there exists a maximum even number N. Then, for every even number n, we have $n \leq N$. Now suppose that $M = N + 2$. Then M is an even integer, since the sum of two even numbers is even by Q1. Since $M = N + 2$, then clearly $M > N$ and so M is even and greater than the largest even number. This contradicts our assumption that $n \leq N$ for all even numbers n. We conclude that there is no maximum even number.

Q3. Suppose, for contradiction, that there exist integers a and b such that $15a+3b = 2$. Dividing by 3, we obtain

$$5a + b = \frac{2}{3}.$$

This is a contradiction, since $5a+b$ is an integer but $\frac{2}{3}$ is not. Hence, we conclude that there are no integers a and b for which $15a + 3b = 2$.

Q4. Suppose, for contradiction, that there exists odd n for which $n^3 + 5$ is odd. Then there exist integers a and b such that $n = 2a + 1$ and $n^3 + 5 = 2b + 1$. Performing some algebraic manipulation,

$$n^3 + 5 = 2b + 1,$$
$$\Rightarrow \quad (2a + 1)^3 + 5 = 2b + 1,$$
$$\Rightarrow \quad 8a^3 + 12a^2 + 6a + 1 + 5 = 2b + 1,$$
$$\Rightarrow \quad 2b - 8a^3 - 12a^2 - 6a = 5,$$
$$\Rightarrow \quad b - 4a^3 - 6a^2 - 3a = \frac{5}{2}.$$

This is a contradiction, since $b - 4a^3 - 6a^2 - 3a$ is an integer but $\frac{5}{2}$ is not. We conclude that if $n^3 + 5$ is odd, then n is even.

Q5. Suppose, for contradiction, that there exists a real number x, with $0 \le x \le \pi/2$, such that $\sin(x) + \cos(x) < 1$. Squaring both sides of the inequality,

$$(\sin(x) + \cos(x))^2 < 1,$$
$$\Rightarrow \quad \sin^2(x) + \cos^2(x) + 2\sin(x)\cos(x) < 1,$$
$$\Rightarrow \quad 1 + 2\sin(x)\cos(x) < 1,$$
$$\Rightarrow \quad 2\sin(x)\cos(x) < 0.$$

For real x, where $0 \le x \le \pi/2$, $\sin(x), \cos(x) \ge 0$. Hence, we have arrived at a contradiction. We conclude that $\sin(x) + \cos(x) \ge 1$ for all real x with $0 \le x \le \pi/2$.

Q6. (a) If a is a factor of b and b is a factor of c then there exist integers k_1 and k_2 such that

$$b = k_1 a, \quad c = k_2 b,$$
$$\Rightarrow \quad c = k_1 k_2 a.$$

Since $k_1 k_2$ is an integer, then a is a factor of c.

(b) Suppose, for contradiction, that a positive integer $n \ge 2$ has second smallest factor m and that m is not prime. If m is not prime, then m has a factor k, where $k < m$. By the result in (a), if k is a factor of m and m is a factor of n, then k is a factor of n. Since $k < m$, then m is *not* the second smallest factor of n, so we have a contradiction. Hence, we conclude that the second smallest factor of a positive integer $n \ge 2$ is prime.

Q7. Suppose, for contradiction, that there exist two positive integers a and b whose sum is non-positive, such that $a + b \leq 0$. Performing some algebraic manipulation,

$$a + b \leq 0,$$
$$\Rightarrow \quad a \leq -b.$$

Since $b > 0$, we have $-b < 0$. However, $a > 0$, so $a \leq -b$ is impossible, so we have arrived at a contradiction. We conclude that the sum of two positive integers is positive.

Q8. Suppose, for contradiction, that there exists a least positive rational number r. However, $\dfrac{r}{2}$ is a positive rational number with $\dfrac{r}{2} < r$, which contradicts r being the smallest positive rational number. Hence, we conclude that there is no least positive rational number.

Q9. Suppose that $a_n \neq 5^n + (-1)^n$ for every positive integer n. Let k be the least positive integer for which the formula fails, i.e. $a_k \neq 5^k + (-1)^k$. Since

$$a_1 = 4 = 5^1 + (-1)^1,$$
$$a_2 = 26 = 5^2 + (-1)^2,$$

the formula is true for $n = 1$ and $n = 2$, so $k \geq 3$. Also, since k is minimal, the formula must be true for $1 \leq l < k$, and since $k \geq 3$, the formula must be true for $k - 1$ and $k - 2$. Beginning with the definition of u_n, we perform some algebraic manipulation,

$$\begin{aligned}
a_k &= 4a_{k-1} + 5a_{k-2}, \quad \text{(by definition of } a_n, \text{ with } n = k) \\
&= 4\left(5^{k-1} + (-1)^{k-1}\right) + 5\left(5^{k-2} + (-1)^{k-2}\right), \\
&= 4 \cdot 5^{k-1} + 4(-1)^{k-1} + 5^{k-1} + 5(-1)^{k-2}, \quad \text{(note that } 5 \cdot 5^{k-2} = 5^{k-1}) \\
&= 5 \cdot 5^{k-1} + 4(-1)^{k-1} + 5(-1)^k, \quad \text{(since } (-1)^{k-2} = (-1)^k) \\
&= 5^k - 4(-1)^k + 5(-1)^k, \quad \text{(since } (-1)^{k-1} = -(-1)^k) \\
&= 5^k + (-1)^k.
\end{aligned}$$

We have in fact deduced that the formula is true for $n = k$, which contradicts our assumption that k is the minimal value for which the formula fails. Since the formula is true for k and for all values of n less than k, and the choice of k is arbitrary, $a_n = 5^n + (-1)^n$ for all positive integer n.

Q10. We use a *minimal criminal* argument here. Suppose, for contradiction, that there exists a positive integer $a \geq 2$ that cannot be written as a product of prime numbers. Moreover, suppose that a is the minimal such positive integer. Clearly, a cannot be prime, so there exist positive integers b and c such that $a = bc$, where $b > 1$ and $c < a$. Both b and c can be written as products of prime numbers, since a is the smallest positive integer that cannot be. But since $a = bc$, this makes a a product of prime numbers, which contradicts our assumption. Since no minimal positive integer exists that cannot be written as a product of prime numbers, we conclude that all positive integers are products of prime numbers.

Q11. Suppose, for contradiction, there exist integers a and b such that $a^2 - 4b = 2$. Hence, $a^2 = 4b + 2 = 2(2b + 1)$, so a^2 is even. Since a^2 is even, it follows that a is even, so $a = 2k$ for some integer k. Thus, k satisfies

$$(2k)^2 - 4b = 2 \Rightarrow 4(k^2 - b) = 2 \Rightarrow 2(k^2 - b) = 1.$$

Since $k^2 - b$ is an integer, we have deduced that 1 is even. Since we know that 1 is not even, we have arrived at a contradiction. Hence, we conclude that our assumption that there exist integers a and b such that $a^2 - 4b = 2$ is false.

Q12. Suppose, for contradiction, that there exist positive integer solutions to the equation $x^2 - y^2 = 1$. Then $(x+y)(x-y) = 1$. It follows that either $x + y = x - y = 1$ or $x + y = x - y = -1$. In the first case, adding the equations gives $x = 1, y = 0$. This contradicts our assumption that y is positive. Similarly, the second case yields $x = -1, y = 0$, again contradicting our assumtion that x and y are positive. Hence, we conclude that there are no integer solutions to the equation $x^2 - y^2 = 1$.

Q13. Suppose, for contradiction, that there exists $n \in \mathbb{N}$ such that

$$\frac{n}{n+1} \geq 1.$$

Since $n + 1 > 0$, we can rearrange the inequality to obtain $n \geq n + 1 \Rightarrow 0 \geq 1$. Clearly, we have arrived at a contradiction, so conclude that out assumption that there exists $n \in \mathbb{N}$ such that $n/(n+1) \geq 1$ is false. Hence, $n/(n+1) < 1$ for all $n \in \mathbb{N}$.

Q14. Suppose, for contradiction, that there exists odd $n \in \mathbb{Z}$ such that $3n + 2$ is even. If n is odd, then there exists $k \in \mathbb{Z}$ such that $n = 2k - 1$. Then

$$3n + 2 = 3(2k - 1) + 2 = 6k - 1 = 2(3k) - 1.$$

Since $3k$ is an integer, we have deduced that $3n + 2$ is odd, contradicting our assumption that $3n + 2$ is even for odd n. Hence, we conclude that if $3n + 2$ is even, n is even.

Q15. Suppose, for contradiction, that there exists even $a \in \mathbb{Z}$ such that $a^2 - 2a + 7$ is even. If a is even, then there exists $k \in \mathbb{Z}$ such that $a = 2k$. Then

$$a^2 - 2a + 7 = (2k)^2 - 2(2k) + 7 = 4k^2 - 4k + 7 = 2(2k^2 - 2k + 3) + 1.$$

Since $2k^2 - 2k + 3$ is an integer, we have deduced that $a^2 - 2a + 7$ is odd, contradicting our assumption that it is even for even a. Hence, we conclude that if $a^2 - 2a + 7$ is even, a is odd.

Q16. Suppose, for contradiction, that there exists $x \in \mathbb{Z}$ such that $2x^3 + 6x + 1 = 0$. Rearranging,

$$1 = -2x^3 - 6x = 2(-x^3 - 3x).$$

Since $-x^3 - 3x$ is an integer, we have deduced that 1 is even. Hence, we have arrived at a contradiction and conclude that $2x^3 + 6x + 1 = 0$ has no integer solutions.

Q17. Suppose, for contradiction, that there exists real $x > 0$ such that

$$x + \frac{1}{x} < 2 \Rightarrow x^2 + 1 < 2x \Rightarrow x^2 - 2x + 1 < 0 \Rightarrow (x - 1)^2 < 0.$$

Since $(x-1)^2 \geq 0$ for all real x, we have arrived at a contradiction. Hence, we conclude that

$$x + \frac{1}{x} \geq 2$$

for all real positive x.

Q18. Suppose, for contradiction, that there exists real $x > 0$ such that

$$\frac{x}{x+1} \geq \frac{x+1}{x+2}.$$

Since x is positive, $x+1$ and $x+2$ are both positive. Hence, we can cross-multiply the fractions without changing the inequality. This yields

$$x(x+2) \geq (x+1)^2 \Rightarrow x^2 + 2x \geq x^2 + 2x + 1 \Rightarrow 0 \geq 1.$$

Hence, we have arrived at a contradiction, so we conclude that

$$\frac{x}{x+1} < \frac{x+1}{x+2}.$$

holds for all real positive x.

Worked Solutions — Exercise 5.3

Q1. Suppose, for contradiction, that there exist rational numbers x and y such that $r = x + y\sqrt{2}$ is rational. Performing some algebraic manipulation,

$$r = \quad x + y\sqrt{2},$$
$$\sqrt{2} = \frac{r-x}{y}.$$

The right-hand side of the expression above is rational (by the algebra of rational numbers). However, $\sqrt{2}$ is irrational, so we have arrived at a contradiction. We conclude that $x + y\sqrt{2}$ is irrational for all rational x and y.

Q2. Suppose, for contradiction, that there exists a rational number x and irrational number y such that $r = x + y$ is rational. Rearranging, we have

$$y = r - x.$$

Since r and x are rational by assumption, $r - x$ is rational by the algebra of rational numbers. However, we assumed that y is rational, so we have arrived at a contradiction. Hence, the sum of a rational number and an irrational number is irrational.

Q3. (a) Suppose, for contradiction, that there exists positive integer n for which n^2 is divisible by 3, but n is not divisible by 3. If n is not divisible by 3, then there exists a positive integer k such that $n = 3k + r$, where $r = 1, 2$. Hence,

$$n^2 = (3k + r)^2,$$
$$= 9k^2 + 6rk + r^2,$$
$$= 3(3k^2rk + 2k) + r^2.$$

Since $r = 1, 2$, $r^2 = 1, 4 \neq 0$. This contradicts our assumption that n^2 is divisible by 3. We conclude that if n^2 is divisible by 3, then n is divisible by 3.

(b) Suppose, for contradiction, that $\sqrt{3}$ is rational. Hence, there exists an integer a and a positive integer b such that

$$\sqrt{3} = \frac{a}{b},$$

where a/b is written in its lowest terms, i.e. the highest common factor of a and b is 1. Performing some algebraic manipulation,

$$\sqrt{3} = \frac{a}{b},$$
$$\Rightarrow \quad 3 = \frac{a^2}{b^2},$$
$$\Rightarrow 3b^2 = a^2.$$

Hence, a^2 is a multiple of 3. Using the result in (a), a is also a multiple of 3, so can be written as $a = 3k$, where k is an integer. Performing further algebraic manipulation,

$$3b^2 = a^2,$$
$$\Rightarrow \quad 3b^2 = 9k^2,$$
$$\Rightarrow \quad b^2 = 3k^2.$$

Hence, b^2 is also a multiple of 3 and so b is a multiple of 3. Since both a and b are multiples of 3, their highest common factor is at least $3 > 1$. This contradicts our assumption that a/b was written in its lowest terms. Since assuming that $\sqrt{3}$ is rational has led to a contradiction, we conclude that $\sqrt{3}$ is irrational.

Q4. (a) Suppose, for contradiction, that $\sqrt{6}$ is rational. Hence, there exists an integer a and a positive integer b such that

$$\sqrt{6} = \frac{a}{b},$$

where a/b is written in its lowest terms, i.e. the highest common factor of a

and b is 1. Performing some algebraic manipulation,

$$\sqrt{6} = \frac{a}{b},$$
$$\Rightarrow \quad 6 = \frac{a^2}{b^2},$$
$$\Rightarrow \quad a^2 = 6b^2,$$

so a^2 is divisible by 6, and so by 2 and 3. From previous results, this means that a is also divisible by 2 and 3, so a is divisible by 6. Say $a = 6k$, where k is an integer. Then

$$a^2 = 36k^2 = 6b^2 \Rightarrow b^2 = 6k^2,$$

so b^2 is divisible by 6. By following the same argument as for a, we conclude that b is also divisible by 6. Hence, the least common factor of a and b is at least $6 > 1$. This contradicts our assumption that a/b was written in its lowest terms. Since assuming that $\sqrt{6}$ is rational has led to a contradiction, we conclude that $\sqrt{6}$ is irrational.

(b) The proof in (a) can be adapted to prove that $\sqrt{8}$ is irrational. We begin again by supposing, for contradiction, that $\sqrt{8}$ is rational, so there exists integer a and positive integer b such that

$$\sqrt{8} = \frac{a}{b},$$

where the highest common factor of a and b is 1. Performing some algebraic manipulation,

$$\sqrt{8} = \frac{a}{b},$$
$$\Rightarrow \quad 8 = \frac{a^2}{b^2}$$
$$\Rightarrow \quad a^2 = 8b^2.$$

This is enough to deduce that a is divisible by 4, but not necessarily by 8. Say $a = 4k$, where k is an integer. Then $a^2 = 16k^2$, and so $b^2 = 2k^2$, which means that b is even. Since a is divisible by 4, it is even. Since both a and b are even, we have a contradiction as in (a). We conclude that $\sqrt{8}$ is irrational.

(c) If we try to adapt the proof in (a) and (b) to prove that $\sqrt{9}$ is irrational, we obtain $a^2 = 9b^2$, so a is divisible by 3, but not necessarily by 9. Say $a = 3k$, where k is an integer. Then $a = 3k^2$, where k is an integer, and so $9k^2 = 9b^2$, and so $k^2 = b^2$. This does not lead to any contradiction, so we cannot proceed with the proof by contradiction. This is not surprising, since $\sqrt{9} = 3$, which is rational.

Q5. (a) We are given that n^3 is even. Suppose, for contradiction, that there exists a that n is odd. Then there exists a positive integer k such that $n = 2k + 1$.

Then

$$n^3 = (2k+1)^3,$$
$$= 8k^3 + 12k^2 + 6k + 1,$$
$$= 2(4k^3 + 6k^2 + 3k) + 1.$$

Since $4k^3 + 6k^2 + 3k$ is an integer, n^3 is odd, which contradicts our premise that n^3 is even. We conclude that if n^3 is even, then n is even.

(b) Suppose, for contradiction, that $2^{1/3}$ is rational. Then there exists an integer a and a positive integer b such that

$$2^{1/3} = \frac{a}{b},$$

where a/b is written in its lowest terms, i.e. the highest common factor of a and b is 1. Performing some algebraic manipulation,

$$2^{1/3} = \frac{a}{b},$$
$$\Rightarrow \quad 2 = \frac{a^3}{b^3},$$
$$\Rightarrow \quad a^3 = 2b^3.$$

Hence, a^3 is even. By the result in (a), a is also even, so can be expressed as $a = 2k$, where k is a positive integer. Performing further algebraic manipulation,

$$a^3 = 2b^3,$$
$$\Rightarrow \quad 8k^3 = 2b^3,$$
$$\Rightarrow \quad 4k^3 = b^3.$$

Hence, b^3 is also even, and so b is even. Since both a and b are even, the highest common factor of a and b is at least $2 > 1$. This contradicts our assumption that a/b was written in its lowest terms. We conclude that $2^{1/3}$ is irrational.

Q6. Given that $ab = c$. Suppose, for contradiction, that $a > \sqrt{c}$ *and* $b > \sqrt{c}$. (Take care with negating the original statement here.) Then

$$ab > \sqrt{c}\sqrt{c} = c.$$

This is a contradiction, since we are given that $ab = c$. We conclude that if $ab = c$, then $a \leq c$ or $b \leq c$.

Q7. Suppose, for contradiction, that the given quadratic equation has a rational solution $x = \frac{p}{q}$, where p is an integer and q is a positive integer, and the highest common factor of p and q is 1. Note that this means that at least one of p and q

is odd. If $\frac{p}{q}$ is indeed a solution to the quadratic, then

$$a\left(\frac{p}{q}\right)^2 + b\left(\frac{p}{q}\right) + c = 0,$$
$$\Rightarrow \quad ap^2 + bpq + cq^2 = 0.$$

The right-hand side of the equation above is even, so the left-hand side must be even also. However, since all of a, b and c are odd, this left-hand side is only even if both p and q are even. This contradicts our assumption that at least one of p and q are odd, so we conclude that the quadratic $ax^2 + bx + c = 0$ has no rational solutions if a, b and c are odd integers.

Q8. Suppose r is a non-zero rational number. Then $r = a/b$ for $a \in \mathbb{Z}$ and $b \in \mathbb{N}$. Also, r can be written as the product of two numbers as follows:

$$r = \sqrt{2} \cdot \frac{r}{\sqrt{2}}.$$

We know (without proof) that $\sqrt{2}$ is irrational. To complete the proof, we must show that $r/\sqrt{2}$ is irrational. Suppose, for contradiction, that $r/\sqrt{2}$ is rational. Then there exist $c \in \mathbb{Z}$, $d \in \mathbb{N}$ such that

$$\frac{r}{\sqrt{2}} = \frac{c}{d}.$$

Then

$$\sqrt{2} = r\frac{d}{c} = \frac{ad}{bc}.$$

Since all of a, b, c, d are integers, we have deduced that $\sqrt{2}$ is rational. Since we know $\sqrt{2}$ is irrational, we have arrived at a contradiction and so conclude that $r/\sqrt{2}$ is irrational. Thus, we can express

$$r = \sqrt{2}\frac{r}{\sqrt{2}}$$

as a product of two irrational numbers.

Q9. The statement is false. For a counter-example, let $a = 2$, $b = 1/2$. Hence, both a and b are rational, but $a^b = 2^{1/2} = \sqrt{2}$ is irrational.

Q10. (a) Suppose, for contradiction, that there is exists an irrational number x such that $x^{1/6}$ is rational. Hence, there exist $a, b \in \mathbb{N}$ such that

$$x^{1/6} = \frac{a}{b} \Rightarrow x = \left(\frac{a}{b}\right)^6 = \frac{a^6}{b^6}.$$

Clearly, a^6 and b^6 are integers, so a^6/b^6 is rational. Hence, we have arrived at a contradiction and so conclude that if x is irrational, then $x^{1/6}$ is also irrational.

(b) The reverse implication of the previous statement is false. As a counter-example, take $x = 8$. Clearly 8 is rational, but

$$8^{1/6} = \left(8^{1/3}\right)^{1/2} = 2^{1/2} = \sqrt{2},$$

so $x^{1/6}$ is irrational. Hence, we have found x such that $x^{1/6}$ is irrational, but x is rational, so the converse is false.

Q11. Note that this activity is challenging and is aimed at high-performing students. Suppose, for contradiction, that $\sqrt{k}$ is not an integer and is rational. Then there exist positive integers a and b such that

$$\sqrt{k} = \frac{a}{b}$$

and the highest common factor of a and b is 1. Let n be the largest integer less than $\sqrt{k}$, so that $0 < \sqrt{k} - n < 1$.

Following the steps given, we obtain

$$\sqrt{k} = \frac{a}{b}$$
$$= \frac{a\left(\sqrt{k} - n\right)}{b\left(\sqrt{k} - n\right)}$$
$$= \frac{a\sqrt{k} - an}{b\sqrt{k} - bn}.$$

Since $\sqrt{k} - n < 1$, the numerator and denominator of the right-hand side are less than a and b respectively. Substitute $a = b\sqrt{k}$ into the numerator, and $\sqrt{k} = a/b$ into the denominator to obtain

$$\sqrt{k} = \frac{b\sqrt{k}\sqrt{k} - an}{b(a/b) - bn}$$
$$= \frac{bk - an}{a - bn}.$$

Recall that a, b, k and n are all positive integers. Also recall that the numerator of the expression on the right-hand side is positive and less than a, and the denominator is positive and less than b. Hence, there exist integers a' and b' such that

$$a' = bk - an < a,$$
$$b' = a - bn \ < b, \text{ and}$$
$$\sqrt{k} = \frac{a'}{b'}.$$

The process above can be repeated indefinitely to produce positive integers a^* and b^* that are successively smaller after each iteration. However, there exists a smallest positive integer, 1. Hence, we arrive at a contradiction and conclude that $\sqrt{k}$ is irrational.

Hence, if $\sqrt{k}$ is not an integer, then it is irrational.

Worked Solutions — Exercise 6.1

Q1. Let $P(n)$ be the statement "$\sum_{r=1}^{n} r = \frac{1}{2}n(n+1)$".

Step 1: We first check the base case, when $n = 1$.

$$LHS = \sum_{r=1}^{1} r$$
$$= 1,$$
$$RHS = \frac{1}{2} \cdot 1 \cdot 2$$
$$= 1.$$

Hence, $P(n)$ is true for the base case $n = 1$.

Step 2: We assume that $P(n)$ is true for $n = k$, that is, we assume that

$$\sum_{r=1}^{k} r = \frac{1}{2}k(k+1)$$

Step 3: We now seek to show that if $P(k)$ is true then $P(k+1)$ must also be true. It often helps to write out the expression we wish to show. In this case, we wish to show that

$$\sum_{r=1}^{k+1} r = \frac{1}{2}(k+1)(k+2)$$

To this end,

$$\sum_{r=1}^{k+1} r = \sum_{r=1}^{k} r + (k+1)$$
$$= \frac{1}{2}k(k+1) + (k+1) \quad \text{by inductive hypothesis}$$
$$= \frac{1}{2}\left[k(k+1) + 2(k+1)\right]$$
$$= \frac{1}{2}(k+1)(k+2)$$

Hence, if the summation formula is true for $n = k$ it is also true for $n = k+1$.

Step 4: Since the summation formula is true for $n = 1$, and if true for $n = k$ then also true for $n = k+1$, it is true for all $n \geq 1$ by the principle of mathematical induction.

Hence, $P(n)$ has been proved $\forall n \in \mathbb{N}$.

Q2. Let $P(n)$ be the statement "$\sum_{r=1}^{n} r^2 = \frac{1}{6}n(n+1)(2n+1)$".

Step 1: We first check the base case, when $n = 1$.

$$LHS = \sum_{r=1}^{1} 1^2$$
$$= 1,$$
$$RHS = \frac{1}{6} \cdot 1 \cdot 2 \cdot 3$$
$$= 1.$$

Hence, $P(n)$ is true for the base case $n = 1$.

Step 2: We assume that $P(n)$ is true for $n = k$, that is, we assume that

$$\sum_{r=1}^{k} r^2 = \frac{1}{6}k(k+1)(2k+1)$$

Step 3: We now seek to show that if $P(k)$ is true then $P(k+1)$ must also be true. It often helps to write out the expression we wish to show. In this case, we wish to show that

$$\sum_{r=1}^{k+1} r^2 = \frac{1}{6}(k+1)(k+2)(2k+3)$$

To this end,

$$\sum_{r=1}^{k+1} r^2 = \sum_{r=1}^{k} r^2 + (k+1)^2$$
$$= \frac{1}{6}k(k+1)(2k+1) + (k+1)^2 \quad \text{by inductive hypothesis}$$
$$= \frac{1}{6}\left[k(k+1)(2k+1) + 6(k+1)^2\right]$$
$$= \frac{1}{6}(k+1)\left[k(2k+1) + 6(k+1)\right]$$
$$= \frac{1}{6}(k+1)\left[2k^2 + k + 6k + 6\right]$$
$$= \frac{1}{6}(k+1)\left[2k^2 + 7k + 6\right]$$
$$= \frac{1}{6}(k+1)(k+2)(2k+3).$$

Hence, if the summation formula is true for $n = k$ it is also true for $n = k+1$.

Step 4: Since the summation formula is true for $n = 1$, and if true for $n = k$ then also true for $n = k+1$, it is true for all $n \geq 1$ by the principle of mathematical induction.

Hence, $P(n)$ has been proved $\forall n \in \mathbb{N}$.

Q3. Let $P(n)$ be the statement "$\sum_{r=1}^{n} r^3 = \frac{1}{4}n^2(n+1)^2$".

Step 1: We first check the base case, when $n = 1$.

$$LHS = \sum_{r=1}^{1} 1^3$$
$$= 1,$$
$$RHS = \frac{1}{4} \cdot 1^2 \cdot 2^2$$
$$= 1.$$

Hence, $P(n)$ is true for the base case $n = 1$.

Step 2: We assume that $P(n)$ is true for $n = k$, that is, we assume that

$$\sum_{r=1}^{k} r^3 = \frac{1}{4} k^2 (k+1)^2$$

Step 3: We now seek to show that if $P(k)$ is true then $P(k+1)$ must also be true. It often helps to write out the expression we wish to show. In this case, we wish to show that

$$\sum_{r=1}^{k+1} r^3 = \frac{1}{4}(k+1)^2(k+2)^2$$

To this end,

$$\sum_{r=1}^{k+1} r^3 = \sum_{r=1}^{k} r^3 + (k+1)^3$$
$$= \frac{1}{4} k^2 (k+1)^2 + (k+1)^3 \quad \text{by inductive hypothesis}$$
$$= \frac{1}{4} \left[k^2 (k+1)^2 + 4(k+1)^3 \right]$$
$$= \frac{1}{4} (k+1)^2 \left[k^2 + 4(k+1) \right]$$
$$= \frac{1}{4} (k+1)^2 \left[k^2 + 4k + 4) \right]$$
$$= \frac{1}{4} (k+1)^2 (k+2)^2.$$

Hence, if the summation formula is true for $n = k$ it is also true for $n = k+1$.

Step 4: Since the summation formula is true for $n = 1$, and if true for $n = k$ then also true for $n = k+1$, it is true for all $n \geq 1$ by the principle of mathematical induction.

Hence, $P(n)$ has been proved $\forall n \in \mathbb{N}$.

Q4. Let $P(n)$ be the statement "$\sum_{r=1}^{n} r(r-1) = \frac{1}{3}(n-1)n(n+1)$".

Step 1: We first check the base case, when $n = 1$.

$$LHS = \sum_{r=1}^{1} 1 \times 0$$
$$= 0,$$
$$RHS = \frac{1}{3} \cdot (1 - 1) \cdot 1 \cdot (1 + 1)$$
$$= \frac{1}{3} \times 0 \times 1 \times 2$$
$$= 0.$$

Hence, $P(n)$ is true for the base case $n = 1$.

Step 2: We assume that $P(n)$ is true for $n = k$, that is, we assume that

$$\sum_{r=1}^{k} r(r - 1) = \frac{1}{3}(k - 1)k(k + 1)$$

Step 3: We now seek to show that if $P(k)$ is true then $P(k + 1)$ must also be true. It often helps to write out the expression we wish to show. In this case, we wish to show that

$$\sum_{r=1}^{k} r(r - 1) = \frac{1}{3}k(k + 1)(k + 2)$$

To this end,

$$\sum_{r=1}^{k+1} r(r - 1) = \sum_{r=1}^{k} r(r - 1) + (k + 1)k$$
$$= \frac{1}{3}(k - 1)k(k + 1) + (k + 1)k \quad \text{by inductive hypothesis}$$
$$= \frac{1}{3}[(k - 1)k(k + 1) + 3(k + 1)k]$$
$$= \frac{1}{3}(k + 1)[(k - 1)k + 3k]$$
$$= \frac{1}{3}k(k + 1)[(k - 1) + 3]$$
$$= \frac{1}{3}k(k + 1)(k + 2)$$

Hence, if the summation formula is true for $n = k$ it is also true for $n = k + 1$.

Step 4: Since the summation formula is true for $n = 1$, and if true for $n = k$ then also true for $n = k + 1$, it is true for all $n \geq 1$ by the principle of mathematical induction.

Hence, $P(n)$ has been proved $\forall n \in \mathbb{N}$.

Q5. Let $P(n)$ be the statement "$\sum_{r=1}^{n} 3^r = \frac{3}{2}(3^n - 1)$".

Step 1: We first check the base case, when $n = 1$.

$$LHS = \sum_{r=1}^{1} 3^1$$
$$= 3,$$
$$RHS = \frac{3}{2}(3^1 - 1)$$
$$= \frac{3}{2} \times 2$$
$$= 3.$$

Hence, $P(n)$ is true for the base case $n = 1$.

Step 2: We assume that $P(n)$ is true for $n = k$, that is, we assume that

$$\sum_{r=1}^{k} 3^r = \frac{3}{2}(3^k - 1)$$

Step 3: We now seek to show that if $P(k)$ is true then $P(k + 1)$ must also be true. It often helps to write out the expression we wish to show. In this case, we wish to show that

$$\sum_{r=1}^{k+1} 3^r = \frac{3}{2}(3^{k+1} - 1)$$

To this end,

$$\sum_{r=1}^{k+1} 3^r = \sum_{r=1}^{k} 3^r + 3^{k+1}$$
$$= \frac{3}{2}(3^k - 1) + 3^{k+1} \quad \text{by inductive hypothesis}$$
$$= \frac{3}{2}\left[3^k - 1 + \frac{2}{3} \times 3^{k+1}\right]$$
$$= \frac{3}{2}\left[3^k - 1 + \frac{2}{3} \times 3 \times 3^k\right]$$
$$= \frac{3}{2}\left[3^k - 1 + 2 \times 3^k\right]$$
$$= \frac{3}{2}(3 \times 3^k - 1)$$
$$= \frac{3}{2}(3^{k+1} - 1).$$

Hence, if the summation formula is true for $n = k$ it is also true for $n = k + 1$.

Step 4: Since the summation formula is true for $n = 1$, and if true for $n = k$ then also true for $n = k + 1$, it is true for all $n \geq 1$ by the principle of mathematical induction.

Hence, $P(n)$ has been proved $\forall n \in \mathbb{N}$.

Q6. Let $P(n)$ be the statement "$\sum_{r=1}^{n} r(r!) = (n+1)! - 1$".

Step 1: We first check the base case, when $n = 1$.

$$LHS = \sum_{r=1}^{1} 1 \times 1!$$
$$= 1,$$
$$RHS = 2! - 1$$
$$= 2 - 1$$
$$= 1.$$

Hence, $P(n)$ is true for the base case $n = 1$.

Step 2: We assume that $P(n)$ is true for $n = k$, that is, we assume that

$$\sum_{r=1}^{k} r(r!) = (k+1)! - 1$$

Step 3: We now seek to show that if $P(k)$ is true then $P(k + 1)$ must also be true. It often helps to write out the expression we wish to show. In this case, we wish to show that

$$\sum_{r=1}^{k+1} r(r!) = (k+2)! - 1$$

To this end,

$$\sum_{r=1}^{k+1} r(r!) = \sum_{r=1}^{k} r(r!) + (k+1)((k+1)!)$$
$$= (k+1)! - 1 + (k+1)(k+1)!$$
$$= (k+1)!(1 + k + 1) - 1$$
$$= (k+2)(k+1)! - 1$$
$$= (k+2)! - 1$$

Hence, if the summation formula is true for $n = k$ it is also true for $n = k + 1$.

Step 4: Since the summation formula is true for $n = 1$, and if true for $n = k$ then also true for $n = k + 1$, it is true for all $n \geq 1$ by the principle of mathematical induction.

Hence, $P(n)$ has been proved $\forall n \in \mathbb{N}$.

Q7. Let $P(n)$ be the statement "$\sum_{r=1}^{n} r^2(r+1) = \frac{1}{12}n(n+1)(n+2)(3n+1)$".

Step 1: We first check the base case, when $n = 1$.

$$LHS = \sum_{r=1}^{1} 1^2 \times 2$$

$$= 2,$$

$$RHS = \frac{1}{12} \times 1 \times 2 \times 3 \times 4$$

$$= \frac{1}{12} \times 24$$

$$= 2.$$

Hence, $P(n)$ is true for the base case $n = 1$.

Step 2: We assume that $P(n)$ is true for $n = k$, that is, we assume that

$$\sum_{r=1}^{k} r^2(r+1) = \frac{1}{12}k(k+1)(k+2)(3k+1)$$

Step 3: We now seek to show that if $P(k)$ is true then $P(k+1)$ must also be true. It often helps to write out the expression we wish to show. In this case, we wish to show that

$$\sum_{r=1}^{k+1} r^2(r+1) = \frac{1}{12}(k+1)(k+2)(k+3)(3(k+1)+1)$$

$$= \frac{1}{12}(k+1)(k+2)(k+3)(3k+4)$$

To this end,

$$\sum_{r=1}^{k+1} r^2(r+1) = \sum_{r=1}^{k} r^2(r+1) + (k+1)^2(k+2)$$

$$= \frac{1}{12}k(k+1)(k+2)(3k+1) + (k+1)^2(k+2)$$

$$= \frac{1}{12}\left[k(k+1)(k+2)(3k+1) + 12(k+1)^2(k+2)\right]$$

$$= \frac{1}{12}(k+1)(k+2)\left[k(3k+1) + 12(k+1)\right]$$

$$= \frac{1}{12}(k+1)(k+2)\left[3k^2 + k + 12k + 12\right]$$

$$= \frac{1}{12}(k+1)(k+2)\left[3k^2 + 13k + 12\right]$$

$$= \frac{1}{12}(k+1)(k+2)(3k+4)(k+3)$$

$$= \frac{1}{12}(k+1)(k+2)(k+3)(3k+4).$$

Hence, if the summation formula is true for $n = k$ it is also true for $n = k + 1$.

Step 4: Since the summation formula is true for $n = 1$, and if true for $n = k$ then also true for $n = k + 1$, it is true for all $n \geq 1$ by the principle of mathematical induction.

Hence, $P(n)$ has been proved $\forall n \in \mathbb{N}$.

Q8. Let $P(n)$ be the statement "$\sum_{r=1}^{n} \frac{2}{(r+1)(r+2)} = \frac{n}{n+2}$".

Step 1: We first check the base case, when $n = 1$.

$$LHS = \sum_{r=1}^{1} \frac{2}{2 \times 3}$$
$$= \frac{1}{3},$$
$$RHS = \frac{1}{1+2} \qquad\qquad = \frac{1}{3}.$$

Hence, $P(n)$ is true for the base case $n = 1$.

Step 2: We assume that $P(n)$ is true for $n = k$, that is, we assume that

$$\sum_{r=1}^{k} \frac{2}{(r+1)(r+2)} = \frac{k}{k+2}.$$

Step 3: We now seek to show that if $P(k)$ is true then $P(k+1)$ must also be true. It often helps to write out the expression we wish to show. In this case, we wish to show that

$$\sum_{r=1}^{k+1} \frac{2}{(r+1)(r+2)} = \frac{k+1}{k+3}$$

To this end,

$$\sum_{r=1}^{k+1} \frac{2}{(r+1)(r+2)} = \sum_{r=1}^{k} \frac{2}{(r+1)(r+2)} + \frac{2}{(k+2)(k+3)}$$
$$= \frac{k}{k+2} + \frac{2}{(k+2)(k+3)}$$
$$= \frac{k(k+3) + 2}{(k+2)(k+3)}$$
$$= \frac{k^2 + 3k + 2}{(k+2)(k+3)}$$
$$= \frac{(k+2)(k+1)}{(k+2)(k+3)}$$
$$= \frac{k+1}{k+3}$$

Hence, if the summation formula is true for $n = k$ it is also true for $n = k + 1$.

Step 4: Since the summation formula is true for $n = 1$, and if true for $n = k$ then also true for $n = k + 1$, it is true for all $n \geq 1$ by the principle of mathematical induction.

Hence, $P(n)$ has been proved $\forall n \in \mathbb{N}$.

Q9. Let $P(n)$ be the statement "$\sum_{r=1}^{2n} \frac{1}{(r+2)(r+3)} = \frac{2n}{6n+9}$".

Step 1: We first check the base case, when $n = 1$.

$$
\begin{aligned}
LHS &= \sum_{r=1}^{2} \frac{1}{3 \times 4} + \frac{1}{20} \\
&= \frac{2}{15}, \\
RHS &= \frac{2}{6+9} \\
&= \frac{2}{15}.
\end{aligned}
$$

Hence, $P(n)$ is true for the base case $n = 1$.

Step 2: We assume that $P(n)$ is true for $n = k$, that is, we assume that

$$
\sum_{r=1}^{2k} \frac{1}{(k+2)(k+3)} = \frac{2k}{6k+9}.
$$

Step 3: We now seek to show that if $P(k)$ is true then $P(k+1)$ must also be true. It often helps to write out the expression we wish to show. In this case, we wish to show that

$$
\begin{aligned}
\sum_{r=1}^{2(k+1)} \frac{1}{(k+2)(k+3)} &= \frac{2(k+1)}{6(k+1)+9} \\
&= \frac{2k+2}{6k+15}
\end{aligned}
$$

To this end,

$$\sum_{r=1}^{2(k+1)} \frac{1}{(k+2)(k+3)} = \sum_{r=1}^{2k} \frac{1}{(k+2)(k+3)} + \frac{1}{(2k+3)(2k+4)}$$

$$+ \frac{1}{(2k+4)(2k+5)}$$

$$= \frac{2k}{6k+9} + \frac{1}{2(2k+3)(k+2)} + \frac{1}{2(k+2)(2k+5)}$$

$$= \frac{2k}{3(2k+3)} \frac{1}{2(2k+3)(k+2)} + \frac{1}{2(k+2)(2k+5)}$$

$$= \frac{4k(k+2)(2k+5) + 3(2k+5) + 3(2k+3)}{6(2k+3)(k+2)(2k+5)}$$

$$= \frac{4k(2k^2+9k+10) + 6k+15 + 6k+9}{6(2k+3)(k+2)(2k+5)}$$

$$= \frac{8k^3 + 36k^2 + 40k + 6k + 15 + 6k + 9}{6(2k+3)(k+2)(2k+5)}$$

$$= \frac{8k^3 + 36k^2 + 52k + 24}{6(2k+3)(k+2)(2k+5)}$$

$$= \frac{4(2k^3 + 9k^2 + 13k + 6)}{6(2k+3)(k+2)(2k+5)}$$

$$= \frac{2(2k+3)(k+1)(k+2)}{3(2k+3)(k+2)(2k+5)}$$

$$= \frac{2(k+1)}{3(2k+5)}$$

$$= \frac{2k+2}{6k+15}.$$

Hence, if the summation formula is true for $n = k$ it is also true for $n = k + 1$.

Step 4: Since the summation formula is true for $n = 1$, and if true for $n = k$ then also true for $n = k + 1$, it is true for all $n \geq 1$ by the principle of mathematical induction.

Hence, $P(n)$ has been proved $\forall n \in \mathbb{N}$.

Q10. Let $P(n)$ be the statement

$$``\sum_{r=1}^{n} \frac{x^{2^{r-1}}}{1 - x^{2^r}} = \frac{1}{1-x} - \frac{1}{1 - x^{2^n}}",$$

where $x \in \mathbb{R}$, such that $x \neq 1$ or $x \neq 1$.

Step 1: We first show that $P(1)$ is true. When $n = 1$,

$$LHS = \frac{x^{2^{1-1}}}{1 - x^{2^1}}$$

$$= \frac{x^{2^0}}{1 - x^{2^1}}$$

$$= \frac{x}{1 - x^2},$$

$$RHS = \frac{1}{1 - x} - \frac{1}{1 - x^{2^1}}$$

$$= \frac{1}{1 - x} - \frac{1}{1 - x^2}$$

$$= \frac{1}{1 - x} - \frac{1}{(1 - x)(1 + x)}$$

$$= \frac{1 + x - 1}{(1 - x)(1 + x)}$$

$$= \frac{x}{1 - x^2}.$$

Hence, $P(n)$ is true for the base case $n = 1$.

Step 2: We assume that $P(n)$ is true for $n = k$, that is, we assume that

$$\sum_{r=1}^{k} \frac{x^{2^{r-1}}}{1 - x^{2^r}} = \frac{1}{1 - x} - \frac{1}{1 - x^{2^k}}.$$

Step 3: We now seek to show that if $P(k)$ is true then $P(k + 1)$ must also be true. It often helps to write out the expression we wish to show. In this case, we wish to show that

$$\sum_{r=1}^{k+1} \frac{x^{2^{r-1}}}{1 - x^{2^r}} = \frac{1}{1 - x} - \frac{1}{1 - x^{2^{k+1}}}$$

To this end, consider,

$$\sum_{r=1}^{k+1} \frac{x^{2^{r-1}}}{1 - x^{2^r}} = \sum_{r=1}^{k} \frac{x^{2^{r-1}}}{1 - x^{2^r}} + \frac{x^{2^k}}{1 - x^{2^{k+1}}}$$

$$= \frac{1}{1 - x} - \frac{1}{1 - x^{2^k}} + \frac{x^{2^k}}{1 - x^{2^{k+1}}}.$$

However,

$$-\frac{1}{1-x^{2^k}}+\frac{x^{2^k}}{1-x^{2^{k+1}}} = -\frac{1}{1-x^{2^k}}+\frac{x^{2^k}}{1-x^{2\cdot 2^k}}$$

$$= -\frac{1}{1-x^{2^k}}+\frac{x^{2^k}}{(1+x^{2^k})(1-x^{2^k})}$$

$$= -\frac{1}{(1+x^{2^k})(1-x^{2^k})}$$

$$= -\frac{1}{1-x^{2^{k+1}}}.$$

Combining these we have,

$$\sum_{r=1}^{k+1}\frac{x^{2^{r-1}}}{1-x^{2^r}} = \frac{1}{1-x}-\frac{1}{1-x^{2^{k+1}}}.$$

Hence, if the summation formula is true for $n=k$ it is also true for $n=k+1$.

Step 4: Since the summation formula is true for $n=1$, and if true for $n=k$ then also true for $n=k+1$, it is true for all $n\geq 1$ by the principle of mathematical induction.

Hence, $P(n)$ has been proved $\forall n\in\mathbb{N}$.

Worked Solutions — Exercise 6.2

Q1. Let $P(n)$ be the statement

$$``\begin{pmatrix}1 & 0\\1 & 1\end{pmatrix}^n = \begin{pmatrix}1 & 0\\n & 1\end{pmatrix}"$$

Step 1: We show $P(1)$ is true. When $n=1$,

$$LHS = \begin{pmatrix}1 & 0\\1 & 1\end{pmatrix},$$

$$RHS = \begin{pmatrix}1 & 0\\1 & 1\end{pmatrix}.$$

Thus, the left hand side is equal to the right hand side and the basis step for the induction argument has been shown.

Step 2: We assume that $P(k)$ is true, that is,

$$\begin{pmatrix}1 & 0\\1 & 1\end{pmatrix}^k = \begin{pmatrix}1 & 0\\k & 1\end{pmatrix}$$

Step 3: Assuming that $P(k)$ is true, we now wish to show that the truth of

$P(k+1)$ immediately follows. We wish to show that,

$$\begin{pmatrix} 1 & 0 \\ 1 & 1 \end{pmatrix}^{k+1} = \begin{pmatrix} 1 & 0 \\ (k+1) & 1 \end{pmatrix}.$$

Consider,

$$\begin{pmatrix} 1 & 0 \\ 1 & 1 \end{pmatrix}^{k+1} = \begin{pmatrix} 1 & 0 \\ 1 & 1 \end{pmatrix}^{k} \begin{pmatrix} 1 & 0 \\ 1 & 1 \end{pmatrix}$$

$$= \begin{pmatrix} 1 & 0 \\ k & 1 \end{pmatrix} \begin{pmatrix} 1 & 0 \\ 1 & 1 \end{pmatrix} \qquad \text{by inductive hypothesis}$$

$$= \begin{pmatrix} 1 \times 1 + 0 \times 1 & 1 \times 0 + 0 \times 1 \\ k \times 1 + 1 \times 1 & k \times 0 + 1 \times 1 \end{pmatrix}$$

$$= \begin{pmatrix} 1 & 0 \\ (k+1) & 1 \end{pmatrix}$$

Hence $P(k+1)$ is also true if $P(k)$ is true.

Step 4: Since the matrix product formula (6.4) is true for $n = 1$, and if true for $n = k$ then also true for $n = k+1$, it is also true for all $n \in \mathbb{N}$ by the principle of mathematical induction.

Q2. Let $P(n)$ be the statement

$$\text{``} \begin{pmatrix} 1 & 0 & 1 \\ 0 & 1 & 0 \\ 0 & 1 & 1 \end{pmatrix}^{n} = \frac{1}{2} \begin{pmatrix} 2 & n^2 - n & 2n \\ 0 & 2 & 0 \\ 0 & 2n & 2 \end{pmatrix} \text{''}$$

Step 1: We show $P(1)$ is true. When $n = 1$,

$$LHS = \begin{pmatrix} 1 & 0 & 1 \\ 0 & 1 & 0 \\ 0 & 1 & 1 \end{pmatrix},$$

$$RHS = \frac{1}{2} \begin{pmatrix} 2 & 1^2 - 1 & 2 \times 1 \\ 0 & 2 & 0 \\ 0 & 2 \times 1 & 2 \end{pmatrix}$$

$$= \begin{pmatrix} 1 & 0 & 1 \\ 0 & 1 & 0 \\ 0 & 1 & 1 \end{pmatrix}.$$

Thus, the left hand side is equal to the right hand side and the basis step for the induction argument has been shown.

Step 2: We assume that $P(k)$ is true, that is,

$$\begin{pmatrix} 1 & 0 & 1 \\ 0 & 1 & 0 \\ 0 & 1 & 1 \end{pmatrix}^k = \frac{1}{2} \begin{pmatrix} 2 & k^2 - k & 2k \\ 0 & 2 & 0 \\ 0 & 2k & 2 \end{pmatrix}$$

Step 3: Assuming that $P(k)$ is true, we now wish to show that the truth of $P(k+1)$ immediately follows. We wish to show that,

$$\begin{pmatrix} 1 & 0 & 1 \\ 0 & 1 & 0 \\ 0 & 1 & 1 \end{pmatrix}^{k+1} = \frac{1}{2} \begin{pmatrix} 2 & (k+1)^2 - (k+1) & 2(k+1) \\ 0 & 2 & 0 \\ 0 & 2(k+1) & 2 \end{pmatrix}.$$

Consider,

$$\begin{pmatrix} 1 & 0 & 1 \\ 0 & 1 & 0 \\ 0 & 1 & 1 \end{pmatrix}^{k+1} = \begin{pmatrix} 1 & 0 & 1 \\ 0 & 1 & 0 \\ 0 & 1 & 1 \end{pmatrix}^k \begin{pmatrix} 1 & 0 & 1 \\ 0 & 1 & 0 \\ 0 & 1 & 1 \end{pmatrix}$$

$$= \frac{1}{2} \begin{pmatrix} 2 & k^2 - k & 2k \\ 0 & 2 & 0 \\ 0 & 2k & 2 \end{pmatrix} \begin{pmatrix} 1 & 0 & 1 \\ 0 & 1 & 0 \\ 0 & 1 & 1 \end{pmatrix}$$

$$= \frac{1}{2} \begin{pmatrix} 2 & k^2 - k + 2k & 2 + 2k \\ 0 & 2 & 0 \\ 0 & 2k + 2 & 2 \end{pmatrix}$$

$$= \frac{1}{2} \begin{pmatrix} 2 & k^2 + k & 2 + 2k \\ 0 & 2 & 0 \\ 0 & 2k + 2 & 2 \end{pmatrix}$$

$$= \frac{1}{2} \begin{pmatrix} 2 & (k+1)^2 - (k+1) & 2(k+1) \\ 0 & 2 & 0 \\ 0 & 2(k+1) & 2 \end{pmatrix}$$

Hence $P(k+1)$ is also true if $P(k)$ is true.

Step 4: Since the matrix product formula (6.4) is true for $n = 1$, and if true for $n = k$ then also true for $n = k+1$, it is also true for all $n \in \mathbb{N}$ by the principle of mathematical induction.

Q3. Let $P(n)$ be the statement

$$\text{``} \begin{pmatrix} \cos(\theta) & \sin(\theta) \\ -\sin(\theta) & \cos(\theta) \end{pmatrix}^n = \begin{pmatrix} \cos(n\theta) & \sin(n\theta) \\ -\sin(n\theta) & \cos(n\theta) \end{pmatrix} \text{''}$$

Step 1: We show $P(1)$ is true. When $n = 1$,

$$LHS = \begin{pmatrix} \cos(\theta) & \sin(\theta) \\ -\sin(\theta) & \cos(\theta) \end{pmatrix},$$

$$RHS = \begin{pmatrix} \cos(n\theta) & \sin(n\theta) \\ -\sin(n\theta) & \cos(n\theta) \end{pmatrix}.$$

Thus, the left hand side is equal to the right hand side and the basis step for the induction argument has been shown.

Step 2: We assume that $P(k)$ is true, that is,

$$\begin{pmatrix} \cos(\theta) & \sin(\theta) \\ -\sin(\theta) & \cos(\theta) \end{pmatrix}^k = \begin{pmatrix} \cos(k\theta) & \sin(k\theta) \\ -\sin(k\theta) & \cos(k\theta) \end{pmatrix}$$

Step 3: Assuming that $P(k)$ is true, we now wish to show that the truth of $P(k+1)$ immediately follows. We wish to show that,

$$\begin{pmatrix} \cos(\theta) & \sin(\theta) \\ -\sin(\theta) & \cos(\theta) \end{pmatrix}^{k+1} = \begin{pmatrix} \cos((k+1)\theta) & \sin((k+1)\theta) \\ -\sin((k+1)\theta) & \cos((k+1)\theta) \end{pmatrix}.$$

Consider,

$$\begin{pmatrix} \cos(\theta) & \sin(\theta) \\ -\sin(\theta) & \cos(\theta) \end{pmatrix}^{k+1} = \begin{pmatrix} \cos(\theta) & \sin(\theta) \\ -\sin(\theta) & \cos(\theta) \end{pmatrix}^k \begin{pmatrix} \cos(\theta) & \sin(\theta) \\ -\sin(\theta) & \cos(\theta) \end{pmatrix}$$

$$= \begin{pmatrix} \cos(k\theta) & \sin(k\theta) \\ -\sin(k\theta) & \cos(k\theta) \end{pmatrix} \begin{pmatrix} \cos(\theta) & \sin(\theta) \\ -\sin(\theta) & \cos(\theta) \end{pmatrix}$$

$$= \begin{pmatrix} B_{11} & B_{12} \\ B_{21} & B_{22} \end{pmatrix},$$

where

$$B_{11} = \cos(k\theta)\cos(\theta) - \sin(k\theta)\sin(\theta)$$
$$B_{12} = \cos(k\theta)\sin(\theta) + \sin(k\theta)\cos(\theta)$$
$$B_{21} = -\sin(k\theta)\cos(\theta) - \cos(k\theta)\sin(\theta)$$
$$B_{22} = -\sin(k\theta)\sin(\theta) + \cos(k\theta)\cos(\theta)$$

Applying the trigonometric addition formula we have,

$$\begin{pmatrix} \cos(\theta) & \sin(\theta) \\ -\sin(\theta) & \cos(\theta) \end{pmatrix}^{k+1} = \begin{pmatrix} \cos((k+1)\theta) & \sin((k+1)\theta) \\ -\sin((k+1)\theta) & \cos((k+1)\theta) \end{pmatrix}$$

Hence $P(k+1)$ is also true if $P(k)$ is true.

Step 4: Since the matrix product formula (6.4) is true for $n = 1$, and if true for $n = k$ then also true for $n = k+1$, it is also true for all $n \in \mathbb{N}$ by the principle of mathematical induction.

Q4. Since we wish to show that $A^n = A$, let $P(n)$ be the statement

$$\text{``}\begin{pmatrix} \cosh^2(\theta) & \cosh^2(\theta) \\ -\sinh^2(\theta) & -\sinh^2(\theta) \end{pmatrix}^n = \begin{pmatrix} \cosh^2(\theta) & \cosh^2(\theta) \\ -\sinh^2(\theta) & -\sinh^2(\theta) \end{pmatrix}\text{''}$$

Step 1: It is clear that $A^1 = A$ and so $P(1)$ is true.

Step 2: We assume that $P(k)$ is true, that is,

$$\begin{pmatrix} \cosh^2(\theta) & \cosh^2(\theta) \\ -\sinh^2(\theta) & -\sinh^2(\theta) \end{pmatrix}^k = \begin{pmatrix} \cosh^2(\theta) & \cosh^2(\theta) \\ -\sinh^2(\theta) & -\sinh^2(\theta) \end{pmatrix}$$

Step 3: Assuming that $P(k)$ is true, we now wish to show that the truth of $P(k+1)$ immediately follows. We wish to show that,

$$\begin{pmatrix} \cosh^2(\theta) & \cosh^2(\theta) \\ -\sinh^2(\theta) & -\sinh^2(\theta) \end{pmatrix}^{k+1} = \begin{pmatrix} \cosh^2(\theta) & \cosh^2(\theta) \\ -\sinh^2(\theta) & -\sinh^2(\theta) \end{pmatrix}.$$

Consider,

$$A^{k+1} = \begin{pmatrix} \cosh^2(\theta) & \cosh^2(\theta) \\ -\sinh^2(\theta) & -\sinh^2(\theta) \end{pmatrix}^k \begin{pmatrix} \cosh^2(\theta) & \cosh^2(\theta) \\ -\sinh^2(\theta) & -\sinh^2(\theta) \end{pmatrix}$$

$$= \begin{pmatrix} \cosh^2(\theta) & \cosh^2(\theta) \\ -\sinh^2(\theta) & -\sinh^2(\theta) \end{pmatrix} \begin{pmatrix} \cosh^2(\theta) & \cosh^2(\theta) \\ -\sinh^2(\theta) & -\sinh^2(\theta) \end{pmatrix}$$

$$= \begin{pmatrix} B_{11} & B_{12} \\ B_{21} & B_{22} \end{pmatrix},$$

where

$$B_{11} = \cosh^2(\theta)(\cosh^2(\theta) - \sinh^2(\theta))$$
$$B_{12} = \cosh^2(\theta)(\cosh^2(\theta) - \sinh^2(\theta))$$
$$B_{21} = -\sinh^2(\theta)(\cosh^2(\theta) - \sinh^2(\theta))$$
$$B_{22} = -\sinh^2(\theta)(\cosh^2(\theta) - \sinh^2(\theta))$$

Since $\cosh^2(\theta) - \sinh^2(\theta) = 1$, we have that $A^{k+1} = A$. Hence $P(k+1)$ is also true if $P(k)$ is true.

Step 4: Since the matrix product formula (6.4) is true for $n = 1$, and if true for $n = k$ then also true for $n = k+1$, it is also true for all $n \in \mathbb{N}$ by the principle of mathematical induction.

Q5. Let $P(n)$ be the statement

$$\text{``}\begin{pmatrix} 10 & -1 \\ 8 & 1 \end{pmatrix}^n = \frac{1}{7}\begin{pmatrix} -2^n + 8 \times 9^n & 2^n - 9^n \\ -8(2^n - 9^n) & 2^{n+3} - 9^n \end{pmatrix}\text{''}$$

Step 1: We show $P(1)$ is true. When $n = 1$,

$$LHS = \begin{pmatrix} 10 & -1 \\ 8 & 1 \end{pmatrix},$$

$$RHS = \frac{1}{7}\begin{pmatrix} -2^1 + 8 \times 9^1 & 2^1 - 9^1 \\ -8(2^1 - 9^1) & 2^{1+3} - 9^1 \end{pmatrix}$$

$$= \frac{1}{7}\begin{pmatrix} 70 & -7 \\ -56 & 7 \end{pmatrix}$$

$$= \begin{pmatrix} 10 & -1 \\ 8 & 1 \end{pmatrix}$$

Thus, the left hand side is equal to the right hand side and the basis step for the induction argument has been shown.

Step 2: We assume that $P(k)$ is true, that is,

$$\begin{pmatrix} 10 & -1 \\ 8 & 1 \end{pmatrix}^k = \frac{1}{7}\begin{pmatrix} -2^k + 8 \times 9^k & 2^k - 9^k \\ -8(2^k - 9^k) & 2^{k+3} - 9^k \end{pmatrix}$$

Step 3: Assuming that $P(k)$ is true, we now wish to show that the truth of $P(k + 1)$ immediately follows. We wish to show that,

$$\begin{pmatrix} 10 & -1 \\ 8 & 1 \end{pmatrix}^{k+1} = \frac{1}{7}\begin{pmatrix} -2^{k+1} + 8 \times 9^{k+1} & 2^{k+1} - 9^{k+1} \\ -8(2^{k+1} - 9^{k+1}) & 2^{k+4} - 9^{k+1} \end{pmatrix}.$$

Consider,

$$\begin{pmatrix} 10 & -1 \\ 8 & 1 \end{pmatrix}^{k+1} = \begin{pmatrix} 10 & -1 \\ 8 & 1 \end{pmatrix}^k \begin{pmatrix} 10 & -1 \\ 8 & 1 \end{pmatrix}$$

$$= \frac{1}{7}\begin{pmatrix} -2^k + 8 \times 9^k & 2^k - 9^k \\ -8(2^k - 9^k) & 2^{k+3} - 9^k \end{pmatrix} \begin{pmatrix} 10 & -1 \\ 8 & 1 \end{pmatrix}$$

$$= \frac{1}{7}\begin{pmatrix} B_{11} & B_{12} \\ B_{21} & B_{22} \end{pmatrix}.$$

Where,

$$B_{11} = 10(-2^k + 8 \times 9^k) + 8(2^k - 9^k)$$
$$= -2 \times 2^k + 72 \times 9^k$$
$$= -2^{k+1} + 8 \times 9^{k+1},$$
$$B_{12} = -(-2^k + 8 \times 9^k) + (2^k - 9^k)$$
$$= 2 \times 2^k - 9 \times 9^k$$
$$= 2^{k+1} - 9^{k+1},$$
$$B_{21} = -80(2^k - 9^k) + 8(2^{k+3} - 9^k)$$
$$= -40 \times 2^{k+1} + 32 \times 2^{k+1} + 72 \times 9^k$$
$$= -8(2^{k+1} - 9^{k+1}),$$
$$B_{22} = 8(2^k - 9^k) + 2^{k+3} - 9^k$$
$$= 2^{k+3} + 2^{k+3} - 9 \times 9^k$$
$$= 2^{k+4} - 9^{k+1}$$

Hence $P(k+1)$ is also true if $P(k)$ is true.

Step 4: Since the matrix product formula (6.4) is true for $n = 1$, and if true for $n = k$ then also true for $n = k+1$, it is also true for all $n \in \mathbb{N}$ by the principle of mathematical induction.

Q6. Let $P(n)$ be the statement

$$\text{``} \begin{pmatrix} 2 & 0 \\ 0 & 2 \end{pmatrix}^n = 2^n \begin{pmatrix} 1 & 0 \\ 0 & 1 \end{pmatrix} \text{''}$$

Step 1: We show $P(1)$ is true. When $n = 1$,

$$LHS = \begin{pmatrix} 2 & 0 \\ 0 & 2 \end{pmatrix},$$

$$RHS = 2^1 \begin{pmatrix} 1 & 0 \\ 0 & 1 \end{pmatrix}.$$

$$= \begin{pmatrix} 2 & 0 \\ 0 & 2 \end{pmatrix}$$

Thus, the left hand side is equal to the right hand side and the basis step for the induction argument has been shown.

Step 2: We assume that $P(k)$ is true, that is,

$$\begin{pmatrix} 2 & 0 \\ 0 & 2 \end{pmatrix}^k = 2^k \begin{pmatrix} 1 & 0 \\ 0 & 1 \end{pmatrix}$$

Step 3: Assuming that $P(k)$ is true, we now wish to show that the truth of $P(k+1)$ immediately follows. We wish to show that,

$$\begin{pmatrix} 2 & 0 \\ 0 & 2 \end{pmatrix}^{k+1} = 2^{k+1} \begin{pmatrix} 1 & 0 \\ 0 & 1 \end{pmatrix}.$$

Consider,

$$\begin{pmatrix} 2 & 0 \\ 0 & 2 \end{pmatrix}^{k+1} = \begin{pmatrix} 2 & 0 \\ 0 & 2 \end{pmatrix}^{k} \begin{pmatrix} 2 & 0 \\ 0 & 2 \end{pmatrix}$$

$$= 2^{k} \begin{pmatrix} 1 & 0 \\ 0 & 1 \end{pmatrix} \begin{pmatrix} 2 & 0 \\ 0 & 2 \end{pmatrix}$$

$$= 2^{k} \begin{pmatrix} 2 & 0 \\ 0 & 2 \end{pmatrix}$$

$$= 2^{k+1} \begin{pmatrix} 1 & 0 \\ 0 & 1 \end{pmatrix}.$$

Hence $P(k+1)$ is also true if $P(k)$ is true.

Step 4: Since the matrix product formula (6.4) is true for $n = 1$, and if true for $n = k$ then also true for $n = k+1$, it is also true for all $n \in \mathbb{N}$ by the principle of mathematical induction.

Q7. (a) Let $P(n)$ be the statement

$$``\begin{pmatrix} 1 & 1 \\ 1 & 1 \end{pmatrix}^{n} = \begin{pmatrix} 2^{n-1} & 2^{n-1} \\ 2^{n-1} & 2^{n-1} \end{pmatrix}"$$

Step 1: We show $P(1)$ is true. When $n = 1$,

$$LHS = \begin{pmatrix} 1 & 1 \\ 1 & 1 \end{pmatrix},$$

$$RHS = \begin{pmatrix} 2^0 & 2^0 \\ 2^0 & 2^0 \end{pmatrix}.$$

$$= \begin{pmatrix} 1 & 1 \\ 1 & 1 \end{pmatrix}$$

Thus, the left hand side is equal to the right hand side and the basis step for the induction argument has been shown.

Step 2: We assume that $P(k)$ is true, that is,

$$\begin{pmatrix} 1 & 1 \\ 1 & 1 \end{pmatrix}^{k} = \begin{pmatrix} 2^{k-1} & 2^{k-1} \\ 2^{k-1} & 2^{k-1} \end{pmatrix}$$

Step 3: Assuming that $P(k)$ is true, we now wish to show that the truth of $P(k+1)$ immediately follows. We wish to show that,

$$\begin{pmatrix} 1 & 1 \\ 1 & 1 \end{pmatrix}^{k+1} = \begin{pmatrix} 2^k & 2^k \\ 2^k & 2^k \end{pmatrix}.$$

Consider,

$$\begin{pmatrix} 1 & 1 \\ 1 & 1 \end{pmatrix}^{k+1} = \begin{pmatrix} 1 & 1 \\ 1 & 1 \end{pmatrix}^{k} \begin{pmatrix} 1 & 1 \\ 1 & 1 \end{pmatrix}$$

$$= \begin{pmatrix} 2^{k-1} & 2^{k-1} \\ 2^{k-1} & 2^{k-1} \end{pmatrix} \begin{pmatrix} 1 & 1 \\ 1 & 1 \end{pmatrix}$$

$$= \begin{pmatrix} 2^{k-1}+2^{k-1} & 2^{k-1}+2^{k-1} \\ 2^{k-1}+2^{k-1} & 2^{k-1}+2^{k-1} \end{pmatrix}$$

$$= \begin{pmatrix} 2 \cdot 2^{k-1} & 2 \cdot 2^{k-1} \\ 2 \cdot 2^{k-1} & 2 \cdot 2^{k-1} \end{pmatrix}$$

$$= \begin{pmatrix} 2^k & 2^k \\ 2^k & 2^k \end{pmatrix}$$

Hence $P(k+1)$ is also true if $P(k)$ is true.

Step 4: Since the matrix product formula is true for $n = 1$, and if true for $n = k$ then also true for $n = k + 1$, it is also true for all $n \in \mathbb{N}$ by the principle of mathematical induction.

(b) Based on the answer to part (a) we conjecture that

$$\begin{pmatrix} 1 & 1 & 1 \\ 1 & 1 & 1 \\ 1 & 1 & 1 \end{pmatrix}^{n} = \begin{pmatrix} 3^{n-1} & 3^{n-1} & 3^{n-1} \\ 3^{n-1} & 3^{n-1} & 3^{n-1} \\ 3^{n-1} & 3^{n-1} & 3^{n-1} \end{pmatrix}$$

We proceed by induction. Let $P(n)$ be the statement

$$\text{``}\begin{pmatrix} 1 & 1 & 1 \\ 1 & 1 & 1 \\ 1 & 1 & 1 \end{pmatrix}^{n} = \begin{pmatrix} 3^{n-1} & 3^{n-1} & 3^{n-1} \\ 3^{n-1} & 3^{n-1} & 3^{n-1} \\ 3^{n-1} & 3^{n-1} & 3^{n-1} \end{pmatrix}.\text{''}$$

Step 1: We show $P(1)$ is true. When $n = 1$,

$$LHS = \begin{pmatrix} 1 & 1 & 1 \\ 1 & 1 & 1 \\ 1 & 1 & 1 \end{pmatrix},$$

$$RHS = \begin{pmatrix} 3^0 & 3^0 & 3^0 \\ 3^0 & 3^0 & 3^0 \\ 3^0 & 3^0 & 3^0 \end{pmatrix}.$$

$$= \begin{pmatrix} 1 & 1 \\ 1 & 1 \end{pmatrix}$$

Thus, the left hand side is equal to the right hand side and the basis step for the induction argument has been shown.

Step 2: We assume that $P(k)$ is true, that is,

$$\begin{pmatrix} 1 & 1 & 1 \\ 1 & 1 & 1 \\ 1 & 1 & 1 \end{pmatrix}^k = \begin{pmatrix} 3^{k-1} & 3^{k-1} & 3^{k-1} \\ 3^{k-1} & 3^{k-1} & 3^{k-1} \\ 3^{k-1} & 3^{k-1} & 3^{k-1} \end{pmatrix}$$

Step 3: Assuming that $P(k)$ is true, we now wish to show that the truth of $P(k+1)$ immediately follows. We wish to show that,

$$\begin{pmatrix} 1 & 1 & 1 \\ 1 & 1 & 1 \\ 1 & 1 & 1 \end{pmatrix}^{k+1} = \begin{pmatrix} 3^k & 3^k & 3^k \\ 3^k & 3^k & 3^k \\ 3^k & 3^k & 3^k \end{pmatrix}.$$

Consider,

$$\begin{pmatrix} 1 & 1 & 1 \\ 1 & 1 & 1 \\ 1 & 1 & 1 \end{pmatrix}^{k+1} = \begin{pmatrix} 1 & 1 & 1 \\ 1 & 1 & 1 \\ 1 & 1 & 1 \end{pmatrix}^k \begin{pmatrix} 1 & 1 & 1 \\ 1 & 1 & 1 \\ 1 & 1 & 1 \end{pmatrix}$$

$$= \begin{pmatrix} 3^{k-1} & 3^{k-1} & 3^{k-1} \\ 3^{k-1} & 3^{k-1} & 3^{k-1} \\ 3^{k-1} & 3^{k-1} & 3^{k-1} \end{pmatrix} \begin{pmatrix} 1 & 1 & 1 \\ 1 & 1 & 1 \\ 1 & 1 & 1 \end{pmatrix}$$

$$= \begin{pmatrix} B & B & B \\ B & B & B \end{pmatrix}.$$

where $B = 3^{k-1} + 3^{k-1} + 3^{k-1}$. Hence,

$$\begin{pmatrix} 1 & 1 & 1 \\ 1 & 1 & 1 \\ 1 & 1 & 1 \end{pmatrix}^{k+1} = \begin{pmatrix} 3 \cdot 3^{k-1} & 3 \cdot 3^{k-1} & 3 \cdot 3^{k-1} \\ 3 \cdot 3^{k-1} & 3 \cdot 3^{k-1} & 3 \cdot 3^{k-1} \\ 3 \cdot 3^{k-1} & 3 \cdot 3^{k-1} & 3 \cdot 3^{k-1} \end{pmatrix}$$

$$= \begin{pmatrix} 3^k & 3^k & 3^k \\ 3^k & 3^k & 3^k \\ 3^k & 3^k & 3^k \end{pmatrix}$$

Hence $P(k+1)$ is also true if $P(k)$ is true.

Step 4: Since the matrix product formula is true for $n = 1$, and if true for $n = k$ then also true for $n = k + 1$, it is also true for all $n \in \mathbb{N}$ by the principle of mathematical induction.

Worked Solutions — Exercise 6.3

Q1. Let $P(n)$ be the statement "$3 \mid 4^n - 1$".

Step 1: When $n = 1$, we have

$$4^1 - 1 = 4 - 1$$
$$= 3,$$

which is certainly divisible by 3 so the statement is true for $n = 1$.

Step 2: We assume the $P(k)$ is true, and so $f(k) = 4^k - 1$ is divisible by 3.

Step 3: We wish to show that $P(k) \Rightarrow P(k + 1)$. We begin with $4^{k+1} - 1$ and attempt to factor out a $4^k - 1$. To this end,

$$4^{k+1} - 1 = 4(4^k) - 1$$
$$= 4(4^k - 1) + 4 - 1$$
$$= 4(4^k - 1) + 3.$$

By our inductive hypothesis the term $4(4^k - 1)$ in the last line above must be divisible by 3. Also, 3 is clearly divisible by 3, and so $4^{k+1} - 1$ must be divisible by 3 as it is the sum of two terms which are each divisible by 3.

Step 4: We have shown the statement to be true for $n = 1$ and if it is true for $n = k$ then it must also be true for $n = k + 1$. Therefore, by the principle of mathematical induction, the statement must be true for all $n \in \mathbb{N}$.

Q2. Let $P(n)$ be the statement "$6 \mid 7^n - 7$".

Step 1: When $n = 1$, we have

$$7^1 - 7 = 0$$

which is certainly divisible by 7 so the statement is true for $n = 1$.

Step 2: We assume the $P(k)$ is true, and so $f(k) = 7^k - 7$ is divisible by 6.

Step 3: We wish to show that $P(k) \Rightarrow P(k + 1)$. We begin with $7^{k+1} - 7$ and attempt to factor out a $7^k - 7$. To this end,

$$7^{k+1} - 7 = 7 \cdot 7^k - 7$$
$$= 7(7^k - 7) + 42$$

By our inductive hypothesis the term $7(7^k - 7)$ in the last line above must be divisible by 6. Also, 42 is clearly divisible by 6, and so $7^{k+1} - 7$ must be divisible by 11 as it is the sum of two terms which are each divisible by 3.

Step 4: We have shown the statement to be true for $n = 1$ and if it is true for $n = k$ then it must also be true for $n = k + 1$. Therefore, by the principle of mathematical induction, the statement must be true for all $n \in \mathbb{N}$.

Q3. Let $P(n)$ be the statement "$3 \mid 5^n + 5$".

Step 1: When $n = 2$, we have

$$5^2 + 5 = 30$$
$$= 5 \times 6,$$

which is certainly divisible by 3 so the statement is true for $n = 1$.

Step 2: We assume the $P(k)$ is true, and so $f(k) = 5^k + 5$ is divisible by 3.

Step 3: We wish to show that $P(k) \Rightarrow P(k + 2)$ since we are only interested in the even integers here. We begin with $5^{k+2} + 5$ and attempt to factor out a $5^k + 5$. To this end,

$$5^{k+2} + 5 = 25 \cdot 5^k + 5$$
$$= 25(5^k + 5) - 120$$

By our inductive hypothesis the term $5(5^k + 5)$ in the last line above must be divisible by 3. Also, -120 is clearly divisible by 3, and so $5^{k+2} + 5$ must be divisible by 3 as it is the sum of two terms which are each divisible by 3.

Step 4: We have shown the statement to be true for $n = 1$ and if it is true for $n = k$ then it must also be true for $n = k + 2$. Therefore, by the principle of mathematical induction, the statement must be true for all $n \in \mathbb{N}$, such that n is even.

Q4. Let $P(n)$ be the statement "$11 \mid 23^n - 1$".

Step 1: When $n = 1$, we have

$$23^1 - 1 = 22$$

which is certainly divisible by 11 so the statement is true for $n = 1$.

Step 2: We assume the $P(k)$ is true, and so $f(k) = 23^k - 1$ is divisible by 11.

Step 3: We wish to show that $P(k) \Rightarrow P(k + 1)$. We begin with $23^{k+1} - 1$ and attempt to factor out a $23^k - 1$. To this end,

$$23^{k+1} - 1 = 23 \cdot 23^k - 1$$
$$= 23(23^k - 1) + 22$$

By our inductive hypothesis the term $23(23^k - 1)$ in the last line above must be divisible by 11. Also, 22 is clearly divisible by 11, and so $23^{k+1} - 1$ must be divisible by 11 as it is the sum of two terms which are each divisible by 3.

Step 4: We have shown the statement to be true for $n = 1$ and if it is true for $n = k$ then it must also be true for $n = k + 1$. Therefore, by the principle of mathematical induction, the statement must be true for all $n \in \mathbb{N}$.

Q5. Let $P(n)$ be the statement "$5 \mid 8^n - 3^n$".

Step 1: When $n = 1$, we have

$$8^1 - 3^1 = 8 - 3$$
$$= 5,$$

which is certainly divisible by 5 so the statement is true for $n = 1$.

Step 2: We assume the $P(k)$ is true, and so $f(k) = 8^k - 3^k$ is divisible by 5.

Step 3: We wish to show that $P(k) \Rightarrow P(k + 1)$. We begin with $8^{k+1} - 3^{k+1}$ and attempt to factor out a $8^k - 3^k$. To this end,

$$8^{k+1} - 3^{k+1} = 8 \cdot 8^k - 3 \cdot 3^k$$
$$= 3 \cdot 8^k + 5 \cdot 8^k - 3 \cdot 3^k$$
$$= 3(8^k - 3^k) + 5 \cdot 8^k$$

By our inductive hypothesis the term $3(8^k - 3^k))$ in the last line above must be divisible by 5. Also, $5 \cdot 8^k$ is clearly divisible by 5, and so $8^{k+1} - 3^{k+1}$ must be divisible by 5 as it is the sum of two terms which are each divisible by 5.

Step 4: We have shown the statement to be true for $n = 1$ and if it is true for $n = k$ then it must also be true for $n = k + 1$. Therefore, by the principle of mathematical induction, the statement must be true for all $n \in \mathbb{N}$.

Q6. Let $P(n)$ be the statement "$5 \mid 3^{3n} + 2^{n+2}$".

Step 1: When $n = 0$, we have

$$3^0 + 2^2 = 1 + 4$$
$$= 5,$$

which is certainly divisible by 5 so the statement is true for $n = 1$.

Step 2: We assume the $P(k)$ is true, and so $f(k) = 3^{3k} + 2^{k+2}$ is divisible by 5.

Step 3: We wish to show that $P(k) \Rightarrow P(k+1)$. We begin with $3^{3(k+1)} + 2^{(k+1)+2}$ and attempt to factor out a $3^{3k} + 2^{k+2}$. To this end,

$$
\begin{aligned}
3^{3(k+1)} + 2^{(k+1)+2} &= 3^{3k+3} + 2^{k+3} \\
&= 3^3 \cdot 3^{3k} + 2 \cdot 2^{k+2} \\
&= 27 \cdot 3^{3k} + 2 \cdot 2^{k+2} \\
&= 2(3^{3k} + 2^{k+2}) + 25 \cdot 3^{3k}
\end{aligned}
$$

By our inductive hypothesis the term $2(3^{3k} + 2^{k+2})$ in the last line above must be divisible by 5. Also, $25 \cdot 3^{3k} = 5 \cdot 5 \cdot 3^{3k}$ is clearly divisible by 5, and so $3^{3(k+1)} + 2^{(k+1)+2}$ must be divisible by 5 as it is the sum of two terms which are each divisible by 5.

Step 4: We have shown the statement to be true for $n = 1$ and if it is true for $n = k$ then it must also be true for $n = k + 1$. Therefore, by the principle of mathematical induction, the statement must be true for all $n \in \mathbb{N}$.

Q7. Let $P(n)$ be the statement "$5 \mid 2^{6n} + 3^{2n-2}$".

Step 1: When $n = 1$, we have

$$2^6 + 3^0 = 65$$

which is certainly divisible by 5 so the statement is true for $n = 1$.

Step 2: We assume the $P(k)$ is true, and so $f(k) = 2^{6k} + 3^{2k-2}$ is divisible by 5.

Step 3: We wish to show that $P(k) \Rightarrow P(k+1)$. We begin with $2^{6(k+1)} + 3^{2(k+1)-2}$

and attempt to factor out a $2^{6k} + 3^{2k-2}$. To this end,

$$2^{6(k+1)} + 3^{2(k+1)-2} = 2^{6k+6} + 3^{2k}$$
$$= 2^6 \cdot 2^{6k} + 3^2 \cdot 3^{2k-2}$$
$$= 9(2^{6k} + 3^{2k-2}) + 55$$

By our inductive hypothesis the term $9(2^{6k} + 3^{2k-2})$ in the last line above must be divisible by 5. Also, 55 is clearly divisible by 5, and so $3^{3(k+1)} + 2^{(k+1)+2}$ must be divisible by 5 as it is the sum of two terms which are each divisible by 5.

Step 4: We have shown the statement to be true for $n = 1$ and if it is true for $n = k$ then it must also be true for $n = k + 1$. Therefore, by the principle of mathematical induction, the statement must be true for all $n \in \mathbb{N}_{\nvdash}$.

Q8. Let $P(n)$ be the statement "$12 \mid 5^{2n+1} + 6^{3n+2} + 7$".

Step 1: When $n = 0$, we have

$$5^1 + 6^2 + 7 = 48$$

which is certainly divisible by 12 so the statement is true for $n = 1$.

Step 2: We assume the $P(k)$ is true, and so $f(k) = 5^{2k+1} + 6^{3k+2} + 7$ is divisible by 12.

Step 3: In this case it is prudent to consider $f(k+1) - f(k)$. If we can show that this quantity is divisible by 12 then $f(k + 1)$ is the sum of two quantities each divisible by 12 and so is also divisible by 12. To this end, we consider,

$$f(k + 1) - f(k) = (5^{2(k+1)+1} + 6^{3(k+1)+2} + 7) - (5^{2k+1} + 6^{3k+2} + 7)$$
$$= (5^{2k+3} + 6^{3k+5} + 7) - (5^{2k+1} + 6^{3k+2} + 7)$$
$$= 25 \cdot 5^{2k+1} + 216 \cdot 6^{3k+2} - 5^{2k+1} - 6^{3k+2}$$
$$= 24 \cdot 5^{2k+1} + 215 \cdot 6^{3k+2}$$
$$= 24 \cdot 5^{2k+1} + 7740 \cdot 6^{3k}$$
$$= 12(2 \cdot 5^{2k+1} + 645 \cdot 6^{3k}),$$

which is a multiple of 12. Hence if $P(k)$ is true then so is $P(k+1)$.

Step 4: We have shown the statement to be true for $n = 1$ and if it is true for $n = k$ then it must also be true for $n = k + 1$. Therefore, by the principle of mathematical induction, the statement must be true for all $n \in \mathbb{N}_{\nvdash}$.

Q9. Let $P(n)$ be the statement "$(x + 1) \mid (x^n + 1)$".

Step 1: Consider the case $n = 1$. The polynomial $x^1 + 1 = x + 1$ is certainly divisible by $(x + 1)$, hence $P(1)$ is true.

Step 2: We assume that $P(k)$ is true. This means that we may write,

$$x^k + 1 = (x+1)p(x),$$

for some polynomial $p(x)$.

Step 3: We wish to show that $(x+1)$ divides $x^{k+1} + 1$. This amounts to showing that,

$$x^{k+1} + 1 = (x+1)q(x),$$

for some polynomial $q(x)$.
To this end,

$$
\begin{aligned}
x^{k+1} - 1 &= x(x^k) - 1 \\
&= x(x^k) - x + x - 1 \\
&= x(x^k + 1) - (x+1) \\
&= x(x+1)p(x) - (x+1) \quad \text{by inductive hypothesis} \\
&= (x+1)(xp(x) - 1).
\end{aligned}
$$

Hence $x^{k+1} + 1 = (x+1)q(x)$ where $q(x) = xp(x) - 1$ and we have shown that if $P(k)$ is true then so is $P(k+1)$.

Step 4: We have shown the statement to be true for $n = 1$ and if it is true for $n = k$ then it must also be true for $n = k + 1$. Therefore, by the principle of mathematical induction, the statement must be true for all $n \in \mathbb{N}$.

Q10. Let $P(n)$ be the statement "$3 \mid 2^{2n} - 1$".

Step 1: When $n = 1$, we have

$$2^{2 \times 1} - 1 = 3$$

which is certainly divisible by 3 so the statement is true for $n = 1$.

Step 2: We assume the $P(k)$ is true, and so $f(k) = 2^{2k} - 1$ is divisible by 3.

Step 3: We wish to show that $P(k) \Rightarrow P(k+1)$. We begin with $2^{2(k+1)} - 1$ and attempt to factor out a $(2^{2k} - 1)$. To this end,

$$
\begin{aligned}
2^{2(k+1)} - 1 &= 2^{2k+2} - 1 \\
&= 2^2 \cdot 2^{2k} - 1 \\
&= 4(2^{2k} - 1) + 3
\end{aligned}
$$

By our inductive hypothesis the term $4(2^{2k} - 1)$ in the last line above must be divisible by 3. Also, 3 is clearly divisible by 3, and so $2^{2(k+1)} - 1$ must be divisible by 3 as it is the sum of two terms which are each divisible by 3.

Step 4: We have shown the statement to be true for $n = 1$ and if it is true for $n = k$ then it must also be true for $n = k + 1$. Therefore, by the principle of mathematical induction, the statement must be true for all $n \in \mathbb{N}_{\nleftarrow}$.

Q11. Let $P(n)$ be the statement "$7 \mid 2^{n+2} + 3^{2n+1}$".

Step 1: When $n = 1$, we have

$$2^{1+2} + 3^{2 \times 1 + 1} = 2^3 + 3^3$$
$$= 35$$

which is certainly divisible by 7 so the statement is true for $n = 1$.

Step 2: We assume the $P(k)$ is true, and so $f(k) = 2^{k+2} + 3^{2k+1}$ is divisible by 7.

Step 3: We wish to show that $P(k) \Rightarrow P(k + 1)$. We begin with $2^{(k+1)+2} + 3^{2(k+1)+1}$ and attempt to factor out a $(2^{k+2} + 3^{2k+1})$. To this end,

$$2^{(k+1)+2} + 3^{2(k+1)+1} = 2^{k+3} + 3^{2k+3}$$
$$= 2 \cdot 2^{k+2} + 9 \cdot 3^{2k+1}$$
$$= 2(2^{k+2} + 3^{2k+1}) + 7 \cdot 3^{2k+1}$$

By our inductive hypothesis the term $2(2^{k+2} + 3^{2k+1})$ in the last line above must be divisible by 7. Also, $7 \cdot 3^{2k+1}$ is clearly divisible by 7, and so $2^{(k+1)+2} + 3^{2(k+1)+1}$ must be divisible by 7 as it is the sum of two terms which are each divisible by 7.

Step 4: We have shown the statement to be true for $n = 1$ and if it is true for $n = k$ then it must also be true for $n = k + 1$. Therefore, by the principle of mathematical induction, the statement must be true for all $n \in \mathbb{N}_{\nleftarrow}$.

Q12. Let $P(n)$ be the statement "$4 \mid 5^n + 3$".

Step 1: When $n = 1$, we have

$$5^1 + 3 = 5 + 3$$
$$= 8,$$

which is certainly divisible by 4 so the statement is true for $n = 1$.

Step 2: We assume the $P(k)$ is true, and so $5^k + 3$ is divisible by 3.

Step 3: We wish to show that $P(k) \Rightarrow P(k + 1)$. We begin with $5^{k+1} + 3$ and attempt to factor out a $5^k + 3$. To this end,

$$5^{k+1} + 3 = 5 \cdot 5^k + 3$$
$$= 5(5^k + 3) - 12$$

By our inductive hypothesis the term $5(5^k + 3)$ in the last line above must be divisible by 4. Also, 12 is clearly divisible by 4, and so $5^{k+1} + 3$ must be divisible by 4 as it is the sum of two terms which are each divisible by 4.

Step 4: We have shown the statement to be true for $n = 1$ and if it is true for $n = k$ then it must also be true for $n = k + 1$. Therefore, by the principle of mathematical induction, the statement must be true for all $n \in \mathbb{N}$.

Worked Solutions — Exercise 6.4

Q1. Let $P(n)$ be the statement "$n! > 15n \quad \forall n \geq 5$".

Step 1: We check the base case, here this is $n = 5$.

$$LHS = 5!$$
$$= 120,$$
$$RHS = 15 * 5$$
$$= 75.$$

Hence, $P(n)$ is true for the base case $n = 5$.

Step 2: We assume that $P(n)$ is true for $n = k$, that is, we assume that

$$k! > 15k.$$

Step 3: We now seek to show that if $P(k)$ is true then $P(k + 1)$ must also be true. It often helps to write out the expression we wish to show. In this case, we wish to show that

$$(k + 1)! > 15(k + 1).$$

To this end,

$$(k + 1)! = (k + 1)k!$$
$$> (k + 1) \times 15k$$
$$> 15k^2 + 15k$$
$$> 15k + 15$$
$$= 15(k + 1)$$

In the second to last line we have used $k > 1$ to conclude that $15k^2 + 15k > 15k + 15$ to obtain the desired result.

Step 4: Since the inequality is true for $n = 5$, and if true for $n = k$ then also true for $n = k + 1$, it is also true for all $n \geq 5$ by the principle of mathematical induction.

Hence, $P(n)$ has been proven for all $n > 5$.

Q2. Let $P(n)$ be the statement "$3^n > 3n \quad \forall n > 1$".

Step 1: We check the base case, here this is $n = 2$.

$$LHS = 3^2$$
$$= 9,$$
$$RHS = 3 \times 2$$
$$= 6.$$

Hence, $P(n)$ is true for the base case $n = 2$.

Step 2: We assume that $P(n)$ is true for $n = k$, that is, we assume that

$$3^k > 3k.$$

Step 3: We now seek to show that if $P(k)$ is true then $P(k+1)$ must also be true. It often helps to write out the expression we wish to show. In this case, we wish to show that

$$3^{k+1} > 3(k+1).$$

To this end,

$$3^{k+1} = 3(3^k)$$
$$> 3(3k),$$
$$> 3(k+1),$$

In the last line we have used $k > 1$ implies that $3k > k+1$ to get the result.

Step 4: Since the inequality is true for $n = 1$, and if true for $n = k$ then also true for $n = k+1$, it is also true for all $n > 1$ by the principle of mathematical induction.

Hence, $P(n)$ has been proven for all $n > 1$.

Q3. Let $P(n)$ be the statement "$5^n + 9 < 6^n \quad \forall n \geq 2$".

Step 1: We check the base case, here this is $n = 2$.

$$LHS = 5^2 + 9$$
$$= 25 + 9$$
$$= 34,$$
$$RHS = 6^2$$
$$= 36$$

Hence, $P(n)$ is true for the base case $n = 2$.

Step 2: We assume that $P(n)$ is true for $n = k$, that is, we assume that

$$5^k + 9 < 6^k.$$

Step 3: We now seek to show that if $P(k)$ is true then $P(k+1)$ must also be true. It often helps to write out the expression we wish to show. In this case, we wish to show that

$$5^{k+1} + 9 < 6^{k+1}.$$

To this end,

$$5^k + 9 < 6^k$$
$$\Rightarrow \qquad 6(5^k + 9) < 6 \times 6^k$$
$$\Rightarrow \qquad (5+1)(5^k + 9) < 6^{k+1}$$
$$\Rightarrow \qquad 5^{k+1} + 45 + 5^k + 9 < 6^{k+1}$$
$$\Rightarrow \quad (5^{k+1} + 9) + (5^k + 45) < 6^{k+1}$$
$$\Rightarrow \qquad 5^{k+1} + 9 < 6^{k+1},$$

since $(5^k + 45) > 0$.

Step 4: Since the inequality is true for $n = 2$, and if true for $n = k$ then also true for $n = k + 1$, it is also true for all $n \geq 2$ by the principle of mathematical induction.

Hence, $P(n)$ has been proven for all $n \geq 2$.

Q4. Let $P(n)$ be the statement "$n^2 \geq 2n + 3 \quad \forall n \geq 1$".

Step 1: We check the base case, here this is $n = 3$.

$$LHS = 3^2$$
$$= 9,$$
$$RHS = 2 \times 3 + 3$$
$$= 9.$$

Hence, $P(n)$ is true for the base case $n = 1$.

Step 2: We assume that $P(n)$ is true for $n = k$, that is, we assume that

$$k^2 \geq 2k + 3$$

Step 3: We now seek to show that if $P(k)$ is true then $P(k+1)$ must also be true. It often helps to write out the expression we wish to show. In this case, we wish to show that

$$(k+1)^2 > 2(k+1) + 3.$$

To this end,

$$(k+1)^2 = k^2 + 2k + 1$$
$$\geq (2k+3) + 2k + 1 \qquad \text{by inductive hypothesis}$$
$$= 4k + 4$$
$$= 2(k+1) + 2k + 2$$
$$\geq 2(k+1) + 3$$

In the last line we have used $k > 3$ implies that $2k + 2 > 3$ to get the result.

Step 4: Since the inequality is true for $n = 3$, and if true for $n = k$ then also true for $n = k+1$, it is also true for all $n \geq 3$ by the principle of mathematical induction.

Hence, $P(n)$ has been proven for all $n \geq 3$.

Q5. Let $P(n)$ be the statement "$n^2 + 8n + 6 < 20n^2 \quad \forall n \geq 1$".

Step 1: We check the base case, here this is $n = 1$.

$$LHS = 1^2 + 8 \times 1 + 6$$
$$= 15,$$
$$RHS = 20 \times 1^2$$
$$= 20.$$

Hence, $P(n)$ is true for the base case $n = 1$.

Step 2: We assume that $P(n)$ is true for $n = k$, that is, we assume that

$$k^2 + 8k + 6 < 20k^2.$$

Step 3: We now seek to show that if $P(k)$ is true then $P(k+1)$ must also be true. It often helps to write out the expression we wish to show. In this case, we wish to show that

$$(k+1)^2 + 8(k+1) + 6 < 20(k+1)^2.$$

To this end,

$$(k+1)^2 + 8(k+1) + 6 = k^2 + 2k + 1 + 8k + 8 + 6$$
$$= (k^2 + 8k + 6) + 2k + 9$$
$$< 20k^2 + 2k + 9$$
$$< 20k^2 + 40k + 9$$
$$< 20k^2 + 40k + 20$$
$$= 20(k^2 + 2k + 1)$$
$$= 20(k+1)^2.$$

Hence if $P(k)$ is true then so is $P(k+1)$.

Step 4: Since the inequality is true for $n = 1$, and if true for $n = k$ then also true for $n = k + 1$, it is also true for all $n \geq 1$ by the principle of mathematical induction.

Q6. Let $P(n)$ be the statement "$(1+x)^n \geq 1 + nx \quad \forall n \geq 1$".

Step 1: We check the base case, here this is $n = 1$.

$$
\begin{aligned}
LHS &= (1+x)^1 \\
&= 1 + x, \\
RHS &= 1 + 1 \times x \\
&= 1 + x.
\end{aligned}
$$

Hence, $P(n)$ is true for the base case $n = 1$.

Step 2: We assume that $P(n)$ is true for $n = k$, that is, we assume that

$$(1+x)^k \geq 1 + kx.$$

Step 3: We now seek to show that if $P(k)$ is true then $P(k+1)$ must also be true. It often helps to write out the expression we wish to show. In this case, we wish to show that

$$(1+x)^{k+1} > 1 + (k+1)x.$$

To this end,

$$
\begin{aligned}
(1+x)^{k+1} &= (1+x)^k(1+x) \\
&> (1+kx)(1+x) \\
&= 1 + x + kx + kx^2 \\
&= (1 + kx + x) + kx^2 \\
&= (1 + (k+1)x) + kx^2 \\
&> 1 + (k+1)x,
\end{aligned}
$$

since $kx^2 > 0$ for all $x \in \mathbb{R}$. Hence if $P(k)$ is true then so is $P(k+1)$.

Step 4: Since the inequality is true for $n = 1$, and if true for $n = k$ then also true for $n = k + 1$, it is also true for all $n \geq 1$ by the principle of mathematical induction.

Q7. Let $P(n)$ be the statement "$12^n > 7^n + 5^n$".

Step 1: We check the base case, here this is $n = 2$.

$$LHS = 12^2$$
$$= 144,$$
$$RHS = 7^2 + 5^2$$
$$= 49 + 25$$
$$= 74$$

Hence, $P(n)$ is true for the base case $n = 2$.

Step 2: We assume that $P(n)$ is true for $n = k$, that is, we assume that

$$12^k > 7^k + 5^k.$$

Step 3: We now seek to show that if $P(k)$ is true then $P(k + 1)$ must also be true. It often helps to write out the expression we wish to show. In this case, we wish to show that

$$12^{k+1} > 7^{k+1} + 5^{k+1}.$$

To this end,

$$12^{k+1} = 12 \times 12^k$$
$$> 12(7^k + 5^k) \quad \text{by inductive hypothesis}$$
$$= 12 \times 7^k + 12 \times 5^k$$
$$> 7 \times 7^k + 12 \times 5^k$$
$$> 7 \times 7^k + 5 \times 5^k$$
$$= 7^{k+1} + 5^{k+1}.$$

Hence if $P(k)$ is true then so is $P(k + 1)$.

Step 4: Since the inequality is true for $n = 2$, and if true for $n = k$ then also true for $n = k + 1$, it is also true for all $n \geq 1$ by the principle of mathematical induction.

Q8. Let $P(n)$ be the statement "$2n < (n + 1)!$".

Step 1: We check the base case, here this is $n = 2$.

$$LHS = 2 \times 2$$
$$= 4,$$
$$RHS = (2 + 1)!$$
$$= 6.$$

Hence, $P(n)$ is true for the base case $n = 2$.

Step 2: We assume that $P(n)$ is true for $n = k$, that is, we assume that

$$2k < (k+1)!.$$

Step 3: We now seek to show that if $P(k)$ is true then $P(k+1)$ must also be true. It often helps to write out the expression we wish to show. In this case, we wish to show that

$$2(k+1) < (k+2)!.$$

To this end,

$$
\begin{aligned}
2(k+1) &= 2k + 2 \\
&< (k+1)! + 2 \\
&< (k+2)!,
\end{aligned}
$$

since $k > 2$. Hence if $P(k)$ is true then so is $P(k+1)$.

Step 4: Since the inequality is true for $n = 2$, and if true for $n = k$ then also true for $n = k + 1$, it is also true for all $n \geq 1$ by the principle of mathematical induction.

Q9. Let $P(n)$ be the statement "$n^3 < 2^n \qquad \forall n \geq 10$".

Step 1: We check the base case, here this is $n = 10$.

$$
\begin{aligned}
LHS &= 10^3 \\
&= 1000, \\
RHS &= 2^{10} \\
&= 1024.
\end{aligned}
$$

Hence, $P(n)$ is true for the base case $n = 10$.

Step 2: We assume that $P(n)$ is true for $n = k$, that is, we assume that

$$k^3 < 2^k.$$

Step 3: We now seek to show that if $P(k)$ is true then $P(k+1)$ must also be true. It often helps to write out the expression we wish to show. In this case, we wish to show that

$$(k+1)^3 < 2^{(k+1)}.$$

To this end,

$$(k+1)^3 = k^3 + 3k^2 + 3k + 1$$
$$< 2^k + 3k^2 + 3k + 1$$
$$< 2^k + 2^k$$
$$= 2 \times 2^k$$
$$= 2^{k+1}$$

since $k > 2$. Hence if $P(k)$ is true then so is $P(k+1)$.

Step 4: Since the inequality is true for $n = 10$, and if true for $n = k$ then also true for $n = k+1$, it is also true for all $n \geq 1$ by the principle of mathematical induction.

Note that in this question we have used $3k^2 + 3k + 1 < 2^k$ which we should really prove by induction too:

Let $P(n)$ be the statement "$3n^2 + 3n + 1 < 2^k$ $\forall n \geq 10$".

Step 1: We check the base case, here this is $n = 10$.

$$LHS = 3 \times 10^2 + 3 \times 10 + 1$$
$$= 331,$$
$$RHS = 2^10$$
$$= 1024.$$

Hence, $P(n)$ is true for the base case $n = 10$.

Step 2: We assume that $P(n)$ is true for $n = k$, that is, we assume that

$$3k^2 + 3k + 1 < 2^k.$$

Step 3: We now seek to show that if $P(k)$ is true then $P(k+1)$ must also be true. It often helps to write out the expression we wish to show. In this case, we wish to show that

$$3(k+1)^2 + 3(k+1) + 1 < 2^k.$$

To this end,

$$3(k+1)^2 + 3(k+1) + 1 = 3k^2 + 6k + 3 + 3k + 3 + 1$$
$$= 3k^2 + 9k + 7$$
$$= (3k^2 + 3k + 1) + 6k + 6$$
$$< 2^k + 6k + 6$$
$$= 2^k + 6(k+1)$$
$$< 2^k + 2^k \qquad\qquad = 2^{k+1}$$

since $k > 2$. Hence if $P(k)$ is true then so is $P(k+1)$.

Step 4: Since the inequality is true for $n = 10$, and if true for $n = k$ then also true for $n = k+1$, it is also true for all $n \geq 1$ by the principle of mathematical induction.

Now note, in this proof we have used that $6k + 6 < 2^k$ which is certainly true when $k > 10$ by monotonicity.

Q10. Let $P(n)$ be the statement "$2n + 2 < 2^n$".

Step 1: We check the base case, here this is $n = 4$.

$$LHS = 2 \times 4 + 2$$
$$= 10,$$
$$RHS = 2^4$$
$$= 16.$$

Hence, $P(n)$ is true for the base case $n = 6$.

Step 2: We assume that $P(n)$ is true for $n = k$, that is, we assume that

$$2k + 2 < 2^k.$$

Step 3: We now seek to show that if $P(k)$ is true then $P(k+1)$ must also be true. It often helps to write out the expression we wish to show. In this case, we wish to show that

$$2(k+1) + 2 < 2^{k+1}.$$

To this end, we consider,

$$2^{k+1} = 2 \times 2^k$$
$$> 2(2k+2)$$
$$= 4k+4$$
$$> 2k+4$$
$$= 2k+2+2$$
$$= 2(k+1)+2$$

since $4k > 2k$. Hence if $P(k)$ is true then so is $P(k+1)$.

Step 4: Since the inequality is true for $n = 2$, and if true for $n = k$ then also true for $n = k + 1$, it is also true for all $n \geq 1$ by the principle of mathematical induction.

Worked Solutions — Exercise 6.5

Q1. Let $P(n)$ be the statement "$u_n = 5 \times 2^{n-1} - 4$".

Step 1: From $u_n = 5 \times 2^{n-1} - 4$ we obtain $u_1 = 5 \times 2^0 - 4 = 1$ which is our given first term. Hence $P(1)$ is true.

Step 2: We assume $P(k)$ holds, that is, we assume that $u_k = 5 \times 2^{k-1} - 4$.

Step 3: For $P(k + 1)$ we wish to show that $u_{k+1} = 5 \times 2^k - 4$; to this end we consider the recurrence formula and apply our inductive hypothesis:

$$u_{k+1} = 2u_k + 4$$
$$= 2(5 \times 2^{k-1} - 4) + 4 \quad \text{by inductive hypothesis}$$
$$= 5 \times 2^k - 8 + 4$$
$$= 5 \times 2^k - 4$$

Hence, if $P(k)$ is true then $P(k+1)$ is also true.

Step 4: Since the result is true for $n = 1$, and if true for $n = k$ it is also true for $n = k + 1$, it is true for all $n \geq 1$ by the principle of mathematical induction.

Q2. Let $P(n)$ be the statement "$u_n = \frac{1}{6}(3^n + 9)$".

Step 1: From $u_n = \frac{1}{6}(3^n + 9)$ we obtain $u_1 = \frac{1}{6}(3^1 + 9) = 2$ which is our given first term. Hence $P(1)$ is true.

Step 2: We assume $P(k)$ holds, that is, we assume that $u_k = \frac{1}{6}(3^k + 9)$.

Step 3: For $P(k + 1)$ we wish to show that $u_{k+1} = \frac{1}{6}(3^{k+1} + 9)$; to this end we consider the recurrence formula and apply our inductive hypothesis:

$$u_{k+1} = 3u_k - 3$$

$$= 3(\frac{1}{6}(3^k + 9)) - 3 \quad \text{by inductive hypothesis}$$

$$= \frac{1}{6}(3^{k+1} + 27 - 18)$$

$$= \frac{1}{6}(3^{k+1} + 9)$$

Hence, if $P(k)$ is true then $P(k+1)$ is also true.

Step 4: Since the result is true for $n = 1$, and if true for $n = k$ it is also true for $n = k + 1$, it is true for all $n \geq 1$ by the principle of mathematical induction.

Q3. Let $P(n)$ be the statement "$u_n = 4^n - 2$".

Step 1: From $u_n = 4^n - 2$ we obtain $u_1 = 4^1 - 2 = 2$ which is our given first term. Hence $P(1)$ is true.

Step 2: We assume $P(k)$ holds, that is, we assume that $u_k = 4^k - 2$.

Step 3: For $P(k + 1)$ we wish to show that $u_{k+1} = 4^{k+1} - 2$; to this end we consider the recurrence formula and apply our inductive hypothesis:

$$u_{k+1} = 4u_k + 6$$

$$= 4(4^k - 2) + 6 \quad \text{by inductive hypothesis}$$

$$= 4^{k+1} - 8 + 6$$

$$= 4^{k+1} - 2.$$

Hence, if $P(k)$ is true then $P(k+1)$ is also true.

Step 4: Since the result is true for $n = 1$, and if true for $n = k$ it is also true for $n = k + 1$, it is true for all $n \geq 1$ by the principle of mathematical induction.

Q4. Let $P(n)$ be the statement "$u_n = \frac{1}{6}(11 \times 3^n - 3)$".

Step 1: From $u_n = \frac{1}{6}(11 \times 3^n - 3)$ we obtain $u_1 = \frac{1}{6}(11 \times 3^1 - 3) = 5$ which is our given first term. Hence $P(1)$ is true.

Step 2: We assume $P(k)$ holds, that is, we assume that $u_k = \frac{1}{6}(11 \times 3^k - 3)$.

Step 3: For $P(k+1)$ we wish to show that $u_{k+1} = \frac{1}{6}(11 \times 3^{k+1} - 3))$; to this end we consider the recurrence formula and apply our inductive hypothesis:

$$u_{k+1} = 3u_k + 1$$
$$= 3(\frac{1}{6}(11 \times 3^k - 3)) + 1 \quad \text{by inductive hypothesis}$$
$$= \frac{1}{6}(11 \times 3^{k+1} - 9 + 6)$$
$$= \frac{1}{6}(11 \times 3^{k+1} - 3)$$

Hence, if $P(k)$ is true then $P(k+1)$ is also true.

Step 4: Since the result is true for $n = 1$, and if true for $n = k$ it is also true for $n = k + 1$, it is true for all $n \geq 1$ by the principle of mathematical induction.

Worked Solutions — Exercise 6.6

Q1. (a) We have,

$$f(x) = \frac{1}{x},$$
$$f'(x) = -1x^{-2},$$
$$f''(x) = 2x^{-3},$$
$$f^{(3)}(x) = -6x^{-4},$$
$$f^{(4)}(x) = 24x^{-5}.$$

(b) We conjecture that for $f(x) = \frac{1}{x}$ we have $f^{(n)}(x) = (-1)^n n! x^{-(n+1)}$ for $n \geq 0$.

(c) We proceed by induction: Let $P(n)$ be the statement

$$\text{“} f^{(n)}(x) = (-1)^n n! x^{-(n+1)} \text{”}$$

Step 1: When $n = 0$ we have,

$$LHS = f^{(0)}(x)$$
$$= f(x)$$
$$= \frac{1}{x},$$
$$RHS = f^{(0)}(x)$$
$$= (-1)^0 0! x^{-(0+1)}$$
$$= \frac{1}{x}$$

Hence $P(n)$ is true for $n = 0$.

Step 2: We assume that $P(k)$ is true,

$$f^{(k)}(x) = (-1)^k k! x^{-(k+1)}$$

Step 3: For $P(k+1)$ we wish to find that

$$f^{(k+1)}(x) = (-1)^{k+1}(k+1)! x^{-((k+1)+1)}$$

Noting that $f^{(k+1)}(x) = \frac{\mathrm{d}}{\mathrm{d}x} f^{(k)}(x)$,

$$
\begin{aligned}
f^{(k+1)}(x) &= \frac{\mathrm{d}}{\mathrm{d}x} f^{(k)}(x) \\
&= \frac{\mathrm{d}}{\mathrm{d}x} (-1)^k k! x^{-(k+1)} \\
&= (-1)^k k! \frac{\mathrm{d}}{\mathrm{d}x} x^{-(k+1)} \\
&= (-1)^k k! \cdot -(k+1) x^{-(k+1)-1} \\
&= (-1)^k (-1)(k+1) k! x^{-((k+1)+1)} \\
&= (-1)^{k+1}(k+1)! x^{-((k+1)+1)}.
\end{aligned}
$$

Step 4: Since $P(n)$ is true for $n = 0$, and if true for $n = k$ then also true for $n = k+1$, it is also true for all $n \geq 0$ by the principle of mathematical induction.

Hence, $P(n)$ has been proven for all $n \in \mathbf{N_0}$.

Q2. Let $P(n)$ be the statement "$(\cosh(\theta) + \sinh(\theta))^n \equiv \cosh(n\theta) + \sinh(n\theta)$". Proceeding by induction:

Step 1: When $n = 1$,

$$
\begin{aligned}
LHS &= (\cosh(\theta) + \sinh(\theta))^1 \\
&= \cosh(\theta) + \sinh(\theta), \\
RHS &= \cosh(1 \times \theta) + \sinh(1 \times \theta) \\
&= \cosh(\theta) + \sinh(\theta).
\end{aligned}
$$

Hence $P(1)$ is true.

Step 2: We assume that $P(k)$ is true,

$$(\cosh(\theta) + \sinh(\theta))^k \equiv \cosh(k\theta) + \sinh(k\theta)$$

Step 3: For $P(k+1)$ we wish to show that

$$(\cosh(\theta) + \sinh(\theta))^{k+1} \equiv \cosh((k+1)\theta) + \sinh((k+1)\theta).$$

To this end,

$$(\cosh(\theta) + \sinh(\theta))^{k+1} \equiv (\cosh(\theta) + \sinh(\theta))^k (\cosh(\theta) + \sinh(\theta))$$
$$\equiv (\cosh(k\theta) + \sinh(k\theta)) (\cosh(\theta) + \sinh(\theta))$$
$$= \cosh(k\theta)\cosh(\theta) + \cosh(k\theta)\sinh(\theta)$$
$$+ \sinh(k\theta)\cosh(\theta) \quad + \sinh(k\theta)\sinh(\theta).$$

However,

$$\cosh(x + y) \equiv \cosh(x)\cosh(y) + \sinh(x)\sinh(y),$$
$$\sinh(x + y) \equiv \sinh(x)\cosh(y) + \cosh(x)\sinh(y).$$

Hence,

$$(\cosh(\theta) + \sinh(\theta))^{k+1} = \cosh((k + 1)\theta) + \sinh((k + 1)\theta),$$

and so if $P(k)$ is true then $P(k + 1)$ is also true.

Step 4: Since $P(n)$ is true for $n = 1$, and if true for $n = k$ then also true for $n = k + 1$, it is also true for all $n \geq 1$ by the principle of mathematical induction.

Q3. Let $P(n)$ be the statement "$\int x^n \, dx = \frac{1}{n+1}x^{n+1} + C$.

Step 1: For $n = 1$, by anti-derivatives,

$$LHS = \int x \, dx$$
$$= \frac{x^2}{2} + C,$$
$$RHS = \frac{1}{1 + 1}x^{1+1} + C$$
$$= \frac{1}{2}x^2 + C.$$

Hence $P(1)$ is true.

Step 2: We assume that $P(k)$ is true, that is,

$$\int x^k \, dx = \frac{1}{k + 1}x^{k+1} + C.$$

Step 3: We wish to show that,

$$\int x^{k+1} \, dx = \frac{1}{k + 2}x^{k+2} + C.$$

Considering $P(k + 1)$,

$$\int x^{k+1} \, dx = \int x \times x^k \, dx.$$

We use integration by parts to evaluate the right hand side of the above equality.

$$\int x \times x^k \; \mathrm{d}x = \frac{x \times x^{k+1}}{k+1} - \int \frac{x^{k+1}}{k+1} \times 1 \; \mathrm{d}x$$

$$\Rightarrow \int x^{k+1} \; \mathrm{d}x = \frac{x \times x^{k+1}}{k+1} - \frac{1}{k+1} \int x^{k+1} \; \mathrm{d}x$$

$$\Rightarrow \left(1 + \frac{1}{k+1}\right) \int x^{k+1} \; \mathrm{d}x = \frac{x^{k+2}}{k+1}$$

$$\Rightarrow \left(\frac{k+2}{k+1}\right) \int x^{k+1} \; \mathrm{d}x = \frac{x^{k+2}}{k+1}$$

$$\Rightarrow \int x^{k+1} \; \mathrm{d}x = \frac{x^{k+2}}{k+2}.$$

So if $P(k)$ is true then $P(k+1)$ is also true.

Step 4: Since $P(n)$ is true for $n = 1$, and if true for $n = k$ then also true for $n = k + 1$, it is also true for all $n \geq 1$ by the principle of mathematical induction.

Q4. Let $P(n)$ be the statement "$\frac{\mathrm{d}^n}{\mathrm{d}\theta^n}(\cos(a\theta)) = a^n \cos\left(a\theta + \frac{n\pi}{2}\right)$".

Step 1: Consider $n = 1$,

$$LHS = \frac{\mathrm{d}}{\mathrm{d}\theta}(\cos(a\theta))$$
$$= -\sin(a\theta)$$
$$= \cos\left(a\theta + \frac{\pi}{2}\right),$$
$$RHS = a^1 \cos\left(a\theta + \frac{1\pi}{2}\right)$$
$$= \left(\cos\left(a\theta + \frac{\pi}{2}\right)\right),$$

where in the third line we have used $-\sin(\theta) = \cos\left(\theta + \frac{\pi}{2}\right)$.

Step 2: We assume that $P(k)$ is true,

$$\frac{\mathrm{d}^k}{\mathrm{d}\theta^k}(\cos(a\theta)) = a^k \cos\left(a\theta + \frac{k\pi}{2}\right).$$

Step 3: We seek to show that $P(k+1)$ follows from $P(k)$,

$$\frac{\mathrm{d}^{k+1}}{\mathrm{d}\theta^{k+1}}(\cos(a\theta)) = a^{k+1} \cos\left(a\theta + \frac{(k+1)\pi}{2}\right).$$

. To this end,

$$\frac{d^{k+1}}{d\theta^{k+1}}(\cos(a\theta)) = \frac{d}{d\theta}\left(\frac{d^k}{d\theta^k}(\cos(a\theta))\right)$$

$$= \frac{d}{d\theta}\left(a^k \cos\left(a\theta + \frac{k\pi}{2}\right)\right) \quad \text{by inductive hypothesis}$$

$$= a^k \frac{d}{d\theta}\left(\cos\left(a\theta + \frac{k\pi}{2}\right)\right)$$

$$= a^k \times -a \sin\left(a\theta + \frac{k\pi}{2}\right)$$

$$= -a^{k+1} \sin\left(a\theta + \frac{k\pi}{2}\right).$$

Recalling that,

$$-\sin(\theta) = \cos\left(\theta + \frac{\pi}{2}\right),$$

we have,

$$\frac{d^{k+1}}{d\theta^{k+1}}(\cos(a\theta)) = a^{k+1} \cos\left(a\theta + \frac{k\pi}{2} + \frac{\pi}{2}\right)$$

$$= a^{k+1} \cos\left(a\theta + \frac{(k+1)\pi}{2}\right).$$

So if $P(k)$ is true then $P(k+1)$ is also true.

Step 4: Since $P(n)$ is true for $n = 1$, and if true for $n = k$ then also true for $n = k + 1$, it is also true for all $n \geq 1$ by the principle of mathematical induction.

Q5. (a) $0, 1, 1, 2, 3, 5, 8, 13$.

(b) We use induction on n.

Let $P(n)$ be the statement,

$$F_n = \frac{\alpha^n - \beta^n}{\sqrt{5}}.$$

Since the recurrence relation for F_n relies on the values of two previous terms, F_{n-1} and F_{n-2} for Step 1 we must check both $n = 0$ and $n = 1$.

Step 1: When $n = 0$,

$$LHS = F_0$$
$$= 0,$$
$$RHS = \frac{\alpha^0 - \beta^0}{\sqrt{5}}$$
$$= \frac{1 - 1}{\sqrt{5}}$$
$$= 0.$$

When $n = 1$,

$$LHS = F_1$$
$$= 1,$$
$$RHS = \frac{\alpha^1 - \beta^1}{\sqrt{5}}$$
$$= \frac{\frac{1+\sqrt{5}}{\sqrt{5}} - \frac{1-\sqrt{5}}{\sqrt{5}}}{\sqrt{5}}$$
$$= \frac{\sqrt{5}}{\sqrt{5}}$$
$$= 1.$$

Hence the result has been shown for $n = 0$ and $n = 1$.

Step 2: We assume that $P(k)$ is true,

$$F_k = \frac{\alpha^k - \beta^k}{\sqrt{5}}.$$

Step 3: We seek to show that if $P(k)$ is true, then $P(k+1)$ is also true,

$$F_{k+1} = \frac{\alpha^{k+1} - \beta^{k+1}}{\sqrt{5}}.$$

To this end, we use the recursive definition of Fibonacci numbers,

$$F_{k+1} = F_k + F_{k-1}$$
$$= \frac{\alpha^k - \beta^k}{\sqrt{5}} + \frac{\alpha^{k-1} - \beta^{k-1}}{\sqrt{5}}$$
$$= \frac{\alpha^{k-1}(\alpha + 1)}{\sqrt{5}} - \frac{\beta^{k-1}(\beta + 1)}{\sqrt{5}}.$$

Now we note that α and β are both roots of the quadratic equation $x^2 - x - 1 =$

0, and so, $x + 1 = x^2$. Using this we obtain,

$$F_{k+1} = \frac{\alpha^{k-1}\alpha^2}{\sqrt{5}} - \frac{\beta^{k-1}\beta^2}{\sqrt{5}}$$

$$= \frac{\alpha^{k+1} - \beta^{k+1}}{\sqrt{5}}.$$

So if $P(k)$ is true then $P(k+1)$ is also true.

Step 4: Since $P(n)$ is true for $n = 0$ and $n = 1$, if true for $n = k$ then also true for $n = k + 1$, and so it is also true for all $n \geq 0$ by the principle of mathematical induction.

Worked Solutions — Exercise 7.1

The converse of Thales' Theorem is that the circumcircle of a right angled triangle has its centre at the midpoint of the hypotenuse of the triangle.

To show this is true, we merely need to show that for a right angled triangle ABC, with right angle at C, the midpoint of the AB (the hypotenuse) is equidistant to all three vertices. We know that the midpoint M of AB is equidistant to A and B, hence we only need to find the distance $|CM|$ and show that it is the same as $|AM|$. Without loss of generality, let us assume that C has coordinates $(0,0)$, A has coordinates $(a,0)$ and B has coordinates $(0,b)$. Using Pythagoras' Theorem, we have the length of the hypotenuse $c = |AB| = \sqrt{a^2 + b^2}$. Now, M has coordinates $(\frac{a}{2}, \frac{b}{2})$ Hence,

$$|CM| = \sqrt{\frac{a^2}{4} + \frac{b^2}{4}} = \frac{1}{2}\sqrt{a^2 + b^2} = \frac{c}{2} = \frac{|AM|}{2}$$

and we are done.

Worked Solutions — Exercise 7.2

The converse of the theroem is: the perpendicular bisector of a chord passes through the centre of the circle.

To prove this, without loss of generality, we assume that the circle is centred at the origin and has radius r. We let $A(x_A, y_A)$ and $B(x_B, y_B)$ be points on the circle, so that $x_A^2 + y_A^2 = r^2$ and $x_B^2 + y_B^2 = r^2$. We consider three cases:

Case 1: $y_A = y_B$.
In this case we must have that x_A and x_B have the same magnitude but opposite signs. The chord points vertically upwards and its midpoint has coordinates $(0, y_B)$. The perpendicular bisector is therefore the y-axis, which must pass through the centre of the circle.

Case 2: $x_A = x_B$.
In this case we must have that y_A and y_B have the same magnitude but opposite signs. The chord has gradient 0 and its midpoint has coordinates $(x_B, 0)$. The perpendicular bisector is therefore the x-axis, which must pass through the centre of the circle.

Case 3: $x_A \neq x_B, y_A \neq y_B$.
In this case, the gradient of the chord is:

$$m_1 = \frac{y_B - y_A}{x_B - x_A}.$$

A line perpendicular to the chord, therefore, has gradient

$$m_2 = \frac{1}{m_1} = \frac{x_B - x_A}{y_B - y_A}.$$

Now, the perpendicular line passing through the midpoint of the chord has equation:

$$y = m_2 x + c,$$

where c is to be determined from the coordinates of the midpoint. If $c = 0$, then we immediately know that the perpendicular bisector passes through the centre of the circle. The midpoint of AB has coordinates $((x_B - x_A)/2, (y_B - y_A)/2$. Hence,

$$\frac{y_B - y_A}{2} = m_2 \left(\frac{x_B - x_A}{2} \right) + c$$

$$= \frac{x_B - x_A}{y_B - y_A} \left(\frac{x_B - x_A}{2} \right) + c.$$

After some manipulation and use of the equation of a circle (omitted for brevity) we find $c = 0$.

Worked Solutions — Exercise 7.3

Consider the triangle ABC in the figure below, where a perpendicular has been dropped from point C. Pythagoras' theorem can be used to find d in both triangles ADC and BDC:

$$d^2 = c^2 - (b + x)^2 \quad \text{and} \quad d^2 = a^2 - x^2.$$

Hence, noting that $x = a \cos(180° - C) = -a \cos(C)$, we find

$$c^2 - (b + x)^2 = a^2 - x^2,$$
$$\Rightarrow \quad c^2 - b^2 - 2bx - x^2 = a^2 - x^2,$$
$$\Rightarrow \quad\quad\quad\quad c^2 = a^2 + b^2 + 2bx,$$
$$\Rightarrow \quad\quad\quad\quad c^2 = a^2 + b^2 - 2ab \cos(C).$$

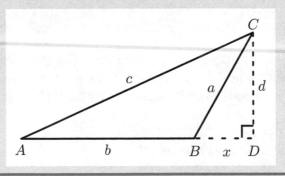

Worked Solutions — Exercise 8.1

From the definition we have

$$f'(x) = \lim_{\delta x \to 0} \frac{\cos(x + \delta x) - \cos(x)}{\delta x}.$$

Applying the addition formula for $\cos(A + B)$ with $A = x$ and $B = \delta x$ gives:

$$f'(x) = \lim_{\delta x \to 0} \frac{\cos(x) \cos(\delta x) - \sin(x) \sin(\delta x) - \cos(x)}{\delta x}.$$

Then, using the small angle approximations, we obtain

$$f'(x) = \lim_{\delta x \to 0} \frac{\cos(x) \left(1 - \frac{1}{2}(\delta x)^2\right) - \sin(x)\delta x - \cos(x)}{\delta x}$$

$$= \lim_{\delta x \to 0} \frac{\cos(x) - \frac{1}{2}(\delta x)^2 \cos(x) - \delta x \sin(x) - \cos(x)}{\delta x}$$

$$= \lim_{\delta x \to 0} \frac{-\delta x \sin(x) - \frac{1}{2}(\delta x)^2 \cos(x)}{\delta x}$$

$$= \lim_{\delta x \to 0} - \sin(x) - \frac{1}{2}\delta x \cos(x).$$

As $\delta x \to 0$, $-\sin(x) - \frac{1}{2}\delta x \cos(x) \to -\sin(x)$ and so $f'(x) = -\sin(x)$.

Worked Solutions — Exercise 8.2

Q1. We shall first consider this function as $y = (u \cdot v) \cdot w$ and apply the product rule.

$$\frac{dy}{dx} = w\frac{d(u \cdot v)}{dx} + (u \cdot v)\frac{dw}{dx}. \tag{11.6}$$

Here we can apply the product rule again on $\frac{d(u \cdot v)}{dx}$ to obtain,

$$\frac{dy}{dx} = w\left(v\frac{du}{dx} + u\frac{dv}{dx}\right) + (u \cdot v)\frac{dw}{dx} \tag{11.7}$$

$$= wu\frac{dv}{dx} + wv\frac{du}{dx} + uv\frac{dw}{dx}. \tag{11.8}$$

Q2. By considering the original product rule and the result of Q1, we make the following *guess* at the formula:

$$g'(x) = \sum_{i=1}^{n} f_i'(x) \prod_{\substack{j \neq i \\ j=1}}^{n} f_j(x).$$

We use induction to prove this: Let $P(k)$ be the statement:

$$\text{If } g(x) = \prod_{i=1}^{k} f_i(x), \text{ then } g'(x) = \sum_{i=1}^{k} f_i'(x) \prod_{\substack{j \neq i \\ j=1}}^{k} f_j(x).$$

Step 1:
$P(2)$ is true immediately due to the product rule with $u(x) = f_1(x)$ and $v(x) = f_2(x)$.

Step 2:
We assume that $P(k)$ is true for some $k \geq 2$.

Step 3:
We show $P(k+1)$ must also be true. Let $g(x) = \prod_{i=1}^{k+1} f_i(x)$ then, we rearrange as follows:

$$g(x) = \underbrace{f_{k+1}(x)}_{u(x)} \underbrace{\prod_{i=1}^{k} f_i(x)}_{v(x)}.$$

Hence $g(x)$ is also a product of two functions $u(x)$ and $v(x)$, which we can differentiate using the product rule and $P(k)$:

$$g'(x) = u'(x)v(x) + u(x)v'(x)$$

$$= f_{k+1}'(x) \prod_{i=1}^{k} f_i(x) + f_{k+1}(x) \sum_{i=1}^{k} f_i'(x) \prod_{\substack{j \neq i \\ j=1}}^{k} f_j(x)$$

$$= f_{k+1}'(x) \prod_{\substack{j \neq k+1 \\ j=1}}^{k+1} f_j(x) + \sum_{i=1}^{k} f_i'(x) \prod_{\substack{j \neq i \\ j=1}}^{k+1} f_j(x)$$

$$= \sum_{i=1}^{k+1} f_i'(x) \prod_{\substack{j \neq i \\ j=1}}^{k+1} f_j(x).$$

Hence, $P(k+1)$ is also true.

Step 4:
As $P(2)$ is true and $P(k) \Rightarrow P(k+1)$, by the principle of mathematical induction, $P(n)$ is true for all $n \geq 2$.

Worked Solutions — Exercise 8.3

Q1. For a function $q(x) = \frac{f(x)}{g(x)}$

$$q'(x) = \lim_{h\to 0} \frac{\frac{f(x+h)}{g(x+h)} - \frac{f(x)}{g(x)}}{h}$$

$$= \lim_{h\to 0} \frac{f(x+h) \cdot g(x) - f(x) \cdot g(x+h)}{g(x) \cdot g(x+h) \cdot h}$$

$$= \lim_{h\to 0} \frac{f(x+h) \cdot g(x) - f(x) \cdot g(x+h) - f(x) \cdot g(x) + f(x) \cdot g(x)}{g(x) \cdot g(x+h) \cdot h}$$

(11.9)

$$= \lim_{h\to 0} \frac{g(x)}{g(x) \cdot g(x+h)} \cdot \frac{f(x+h) - f(x)}{h}$$

$$- \lim_{h\to 0} \frac{f(x)}{g(x) \cdot g(x+h)} \cdot \frac{g(x+h) - g(x)}{h}$$

$$= \frac{g(x) \cdot f'(x) - f(x) \cdot g'(x)}{(g(x))^2}.$$

Similar to the product rule, we add and subtract the term $f(x) \cdot g(x)$ in (11.9) effectively adding zero in order to factorise the expression.

Q2. (a) We recall

$$\tan(x) = \frac{\sin(x)}{\cos(x)}.$$

Hence, we apply the quotient rule with $u = \sin(x)$ and $v = \cos(x)$. Then

$$\frac{\mathrm{d}\tan(x)}{\mathrm{d}x} = \frac{\cos(x) \cdot \cos(x) - \sin(x) \cdot \sin(x)}{\cos^2(x)}$$

$$= \frac{\cos^2(x) + \sin^2(x)}{\cos^2(x)}$$

$$= \frac{1}{\cos^2(x)}$$

$$= \sec^2(x),$$

where we have used the trigonometric identity $\cos^2(x) + \sin^2(x) \equiv 1$.

(b) We recall

$$\sec(x) = \frac{1}{\cos(x)}.$$

Hence, we apply the quotient rule with $u = 1$ and $v = \cos(x)$. Then

$$\frac{\mathrm{d}\sec(x)}{\mathrm{d}x} = \frac{\cos(x) \cdot 0 + 1 \cdot \sin(x)}{\cos^2(x)}$$

$$= \frac{\sin(x)}{\cos^2(x)}$$

$$= \tan(x)\sec(x).$$

(c) We recall

$$\operatorname{cosec}(x) = \frac{1}{\sin(x)}.$$

Hence, we apply the quotient rule with $u = 1$ and $v = \sin(x)$. Then

$$\frac{d \operatorname{cosec}(x)}{dx} = \frac{\sin(x) \cdot 0 - 1 \cdot \cos(x)}{\sin^2(x)}$$

$$= \frac{-\cos(x)}{\sin^2(x)}$$

$$= -\cot(x)\operatorname{cosec}(x).$$

Worked Solutions — Exercise 8.4

Q1. We let $y = \frac{1}{v}$, with v a function of x, then the chain rule gives:

$$\frac{dy}{dx} = \frac{dy}{dv} \cdot \frac{dv}{dx}$$

$$= -\frac{1}{v^2} \cdot v'(x)$$

$$= -\frac{v'(x)}{v^2(x)},$$

as required.

Q2. Let $g(x) = \frac{u(x)}{v(x)}$, then, letting $y(x) = \frac{1}{v(x)}$, we have $g(x) = u(x)y(x)$. Applying the product rule and the result of Q1, we find

$$\frac{dg}{dx} = u\frac{dy}{dx} + y\frac{du}{dx}$$

$$= -u\frac{\frac{dv}{dx}}{v^2(x)} + \frac{1}{v} \cdot \frac{du}{dx}$$

$$= \frac{v\frac{du}{dx} - u\frac{du}{dx}}{v^2(x)},$$

which is the quotient rule, as required.

Q3. In this question, we make use of the result that if x is measured in radians and θ is measured in degrees, then $x(\theta) = \frac{\theta\pi}{180}$. Let $\widehat{\sin}(x)$ be the radian version of sin, then $\widehat{\sin}(x) = \sin(\theta)$, if $x = \frac{\theta\pi}{180}$. Let us use similar notation for the other trigonometric functions in radians.

(a) We use the chain rule as follows:

$$\frac{d \sin(\theta)}{d\theta} = \frac{d\widehat{\sin}(x)}{dx} \cdot \frac{dx}{d\theta}$$

$$= \widehat{\cos}(x) \cdot \frac{\pi}{180}$$

$$= \frac{\pi}{180}\cos(\theta)$$

(b) We repeat the same technique for $\cos(\theta)$:

$$\frac{\mathrm{d}\cos(\theta)}{\mathrm{d}\theta} = \frac{\mathrm{d}\widehat{\cos}(x)}{\mathrm{d}x} \cdot \frac{\mathrm{d}x}{\mathrm{d}\theta}$$
$$= -\widehat{\sin}(x) \cdot \frac{\pi}{180}$$
$$= -\frac{\pi}{180}\sin(\theta)$$

(c) We repeat the same technique for $\tan(\theta)$:

$$\frac{\mathrm{d}\tan}{\mathrm{d}\theta} = \frac{\mathrm{d}\widehat{\tan}(x)}{\mathrm{d}x} \cdot \frac{\mathrm{d}x}{\mathrm{d}\theta}$$
$$= -\widehat{\sec}^2(x) \cdot \frac{\pi}{180}$$
$$= -\frac{\pi}{180}\sec^2(\theta)$$

Worked Solutions — Exercise 8.5

First we define a function $G(x)$ that is continuous on the interval $[a, b]$ such that

$$G(x) = \int_a^x F'(t)\,\mathrm{d}t.$$

Then, by the Fundamental Theorem of Calculus Part I,

$$G'(x) = F'(x)$$

If the derivatives are equal, then integrating each side would give us

$$G(x) + c_1 = F(x) + c_2.$$

Therefore, $G(x)$ and $F(x)$ differ by a constant, c. In other words,

$$G(x) + c = F(x)$$

for all $x \in [a, b]$. Therefore, taking the right hand side

$$F(b) - F(a) = (G(b) + c) - (G(a) + c) = G(b) - G(a) = \int_a^b F'(t)\,\mathrm{d}t - \int_a^a F'(t)\,\mathrm{d}t.$$

We know that

$$\int_a^a F'(t)\,\mathrm{d}t = 0,$$

and so,

$$F(b) - F(a) = \int_a^b F'(t)\,\mathrm{d}t$$

as required.

Worked Solutions — Exercise 8.6

This proof uses Theorem 8.13 and the result on the linearity of differentiation Theorem 8.5. Suppose that $F(x)$ and $G(x)$ are the antiderivatives of $f(x)$ and $g(x)$, respectively. Then

$$\frac{\mathrm{d}}{\mathrm{d}x}[F(x) + G(x)] = \frac{\mathrm{d}F(x)}{\mathrm{d}x} + \frac{\mathrm{d}G(x)}{\mathrm{d}x} = g(x) + f(x).$$

Thus, by the Fundamental Theorem of Calculus Part II (twice) we obtain

$$\int_a^b f(x) + g(x)\,\mathrm{d}x = [F(x) + G(x)]_a^b$$
$$= F(b) + G(b) - (F(a) + G(a))$$
$$= (F(b) - F(a)) + (G(b) - G(a))$$
$$= \int_a^b f(x)\,\mathrm{d}x + \int_a^b g(x)\,\mathrm{d}x.$$

Similarly,

$$\frac{\mathrm{d}}{\mathrm{d}x}[F(x) - G(x)] = \frac{\mathrm{d}F(x)}{\mathrm{d}x} - \frac{\mathrm{d}G(x)}{\mathrm{d}x} = g(x) - f(x).$$

Thus, by the Fundamental Theorem of Calculus Part II (twice) we obtain

$$\int_a^b f(x) - g(x)\,\mathrm{d}x = [F(x) - G(x)]_a^b$$
$$= F(b) - G(b) - (F(a) - G(a))$$
$$= (F(b) - F(a)) - (G(b) - G(a))$$
$$= \int_a^b f(x)\,\mathrm{d}x - \int_a^b g(x)\,\mathrm{d}x$$
$$= \int_a^b f(x)\,\mathrm{d}x + \int_b^a g(x)\,\mathrm{d}x.$$

Worked Solutions — Exercise 8.7

Q1. Let $F(x) = u(x)v(x)$, then the product rule gives

$$\frac{\mathrm{d}F}{\mathrm{d}x} = \frac{\mathrm{d}(uv)}{\mathrm{d}x} = u\frac{\mathrm{d}v}{\mathrm{d}x} + v\frac{\mathrm{d}u}{\mathrm{d}x}.$$

The left hand side is a perfect derivative, so we can integrate both sides and apply the second part of the Fundamental Theorem of Calculus to obtain

$$[uv]_a^b = \int_a^b \left(u \frac{dv}{dx} + v \frac{du}{dx} \right) dx$$

$$= \int_a^b u \frac{dv}{dx}\, dx + \int_a^b v \frac{du}{dx}\, dx,$$

$$\Rightarrow \quad [uv]_a^b - \int_a^b v \frac{du}{dx}\, dx = \int_a^b u \frac{dv}{dx}\, dx.$$

Hence, the integration by parts formula is proved.

Q2. Let $F = F(u)$ be the antiderivative of $f(u)$, with respect to u and let u be a function of x. Then, the composite function $F(u(x))$ exists and we can apply the chain rule to it to find

$$\frac{dF(u(x))}{dx} = \frac{dF}{du} \frac{du}{dx}$$

$$= f(u(x)) \frac{du}{dx}.$$

Now, we integrate both sides of the above with respect to x between a and b and use the Fundamental Theorem of Calculus Part II to obtain

$$\int_a^b f(u(x)) \frac{du}{dx}\, dx = \int_a^b \frac{dF(u(x))}{dx}\, dx$$

$$= [F(u(x)]_a^b$$

$$= F(u(b)) - F(u(a)).$$

As $F(u)$ is the antiderivative of $f(u)$, then we can use the Fundamental Theorem of Calculus Part II the other way, so that:

$$F(u(b)) - F(u(a)) = \int_{u(a)}^{u(b)} f(u)\, du.$$

Combining these results gives:

$$\int_a^b f(u(x)) \frac{du}{dx}\, dx = \int_{u(a)}^{u(b)} f(u)\, du.$$

Worked Solutions — Exercise 9.1

Q1.

$$\frac{\sin^2(\theta)}{\tan^2(\theta)} = \frac{\sin^2(\theta)}{\frac{\sin^2(\theta)}{\cos^2(\theta)}},$$

$$= \frac{\sin^2(\theta)\cos^2(\theta)}{\sin^2(\theta)},$$

$$= \cos^2(\theta),$$

$$= 1 - \sin^2(\theta).$$

Q2.

$$f(x) = \tan^2(x)\cos^4(x),$$

$$= \frac{\sin^2(x)}{\cos^2(x)}\cos^4(x),$$

$$= \sin^2(x)\cos^2(x),$$

$$= \sin^2(x)(1 - \sin^2(x)),$$

$$= \sin^2(x) - \sin^4(x).$$

Q3. First we note the following,

$$\cos(\theta)\tan(\theta) = \cos(\theta)\frac{\sin(\theta)}{\cos(\theta)},$$

$$= \sin(\theta).$$

Hence,

$$\cos(\theta)\tan(\theta) = \frac{\sqrt{3}}{2},$$

$$\Rightarrow \qquad \sin(\theta) = \frac{\sqrt{3}}{2}.$$

Thus, $\theta = 60°, 120°$.

Q4. We first seek to simplify the left hand side of the equation we need to solve. To this end, we apply the trigonometric identity (9.2),

$$\frac{1 - 2\cos^2(\theta) + \cos^4(\theta)}{\sin^2(\theta)} = \frac{(1 - \cos^2(\theta))^2}{\sin^2(\theta)},$$

$$= \frac{(1 - \cos^2(\theta))^2}{1 - \cos^2(\theta)},$$

$$= 1 - \cos^2(\theta).$$

Using the above,

$$\frac{1 - 2\cos^2(\theta) + \cos^4(\theta)}{\sin^2(\theta)} = \frac{1}{2},$$

$$\Rightarrow \qquad 1 - \cos^2(\theta) = \frac{1}{2},$$

$$\Rightarrow \qquad \cos^2(\theta) = \frac{1}{2},$$

$$\Rightarrow \qquad \cos(\theta) = \frac{1}{\sqrt{2}}.$$

Using the periodicity of $\cos(\theta)$ we see that $\theta \in \{45°, 315°\}$.

Q5. (a) Starting from the left hand side and applying trigonometric identity (9.2),

$$\frac{\cos^4(\theta) - \sin^4(\theta)}{\cos^2(\theta)} = \frac{\left(\cos^2(\theta) + \sin^2(\theta)\right)\left(\cos^2(\theta) - \sin^2(\theta)\right)}{\cos^2(\theta)},$$

$$= \frac{\cos^2(\theta) - \sin^2(\theta)}{\cos^2(\theta)},$$

$$= \frac{\cos^2(\theta)}{\cos^2(\theta)} - \frac{\sin^2(\theta)}{\cos^2(\theta)},$$

$$= 1 - \tan^2(\theta).$$

(b) Using the result shown above,

$$\frac{\cos^4(\theta) - \sin^4(\theta)}{\cos^2(\theta)} = \frac{1}{2},$$

$$\Rightarrow \qquad 1 - \tan^2(\theta) = \frac{1}{2},$$

$$\Rightarrow \qquad \tan^2(\theta) = \frac{1}{2},$$

$$\Rightarrow \qquad \tan(\theta) = \frac{1}{\sqrt{2}}.$$

As $\tan(\theta)$ has period $180°$, there are three solutions in the range $0° \leq \theta \leq 540°$, namely $\theta = 35.26°, 215.26°, 395.26°$.

Q6. We first consider,

$$(\cos(x) + \sin(x))^3 = \cos^3(x) + 3\sin(x)\cos^2(x) + 3\sin^2(x)\cos(x) + \sin^3(x),$$

$$= \cos^3(x) + 3\sin(x)\cos^2(x) + 3(1 - \cos^2(x))\cos(x) + \sin(x)(1 - \cos^2(x)),$$

$$= \cos^3(x) + 3\sin(x)\cos^2(x) + 3\cos(x) - 3\cos^3(x) + \sin(x) - \sin(x)\cos^2(x),$$

$$= 3\cos(x) - 2\cos^3(x) + 2\sin(x)\cos^2(x) + \sin(x).$$

Hence,

$$f(x) = (\cos(x) + \sin(x))^3 - \sin(x)\left(2\cos^2(x) + 1\right),$$
$$= 3\cos(x) - 2\cos^3(x) + 2\sin(x)\cos^2(x) + \sin(x)$$
$$\quad - \sin(x)\left(2\cos^2(x) + 1\right),$$
$$= 3\cos(x) - 2\cos^3(x) + 2\sin(x)\cos^2(x) + \sin(x)$$
$$\quad - 2\sin(x)\cos^2(x) - \sin(x),$$
$$= 3\cos(x) - 2\cos^3(x).$$

Q7. (a) This particular identity will reappear when we study the derivatives of trigonometric functions. We have

$$\frac{\sin(x)}{\cos^2(x)} = \frac{1}{\cos(x)} \times \frac{\sin(x)}{\cos(x)}$$
$$= \sec(x)\tan(x).$$

(b) We rewrite the left hand side in terms of sine and cosine:

$$\tan(x) + \cot(x) = \frac{\sin(x)}{\cos(x)} + \frac{\cos(x)}{\sin(x)}$$
$$= \frac{\sin^2(x)}{\sin(x)\cos(x)} + \frac{\cos^2(x)}{\sin(x)\cos(x)}$$
$$= \frac{\sin^2(x) + \cos^2(x)}{\sin(x)\cos(x)}$$
$$= \frac{1}{\sin(x)\cos(x)} \qquad \text{(Pythagorean identity)}$$
$$= \frac{1}{\cos(x)} \times \frac{1}{\sin(x)}$$
$$= \sec(x)\cosec(x).$$

(c) There does not seem much alternative beyond writing things in terms of sine and cosine.

The left-hand side is

$$\cosec(x) + \tan(x)\sec(x) = \frac{1}{\sin(x)} + \frac{\sin(x)}{\cos(x)} \times \frac{1}{\cos(x)}$$
$$= \frac{\cos^2(x)}{\sin(x)\cos^2(x)} + \frac{\sin^2(x)}{\sin(x)\cos^2(x)}$$
$$= \frac{\cos^2(x) + \sin^2(x)}{\sin(x)\cos^2(x)}$$
$$= \frac{1}{\sin(x)\cos^2(x)}$$

using the Pythagorean identity.

Meanwhile, the right hand side is

$$\operatorname{cosec}(x)\sec^2(x) = \frac{1}{\sin(x)} \times \frac{1}{\cos^2(x)},$$

which is the same, and so

$$\operatorname{cosec}(x) + \tan(x)\sec(x) = \operatorname{cosec}(x)\sec^2(x).$$

(d) We tackle the two identities separately. The first identity makes use of the difference of two squares:

$$\sec^4(x) - \tan^4(x) = (\sec^2(x) + \tan^2(x))(\sec^2(x) - \tan^2(x))$$
$$= \sec^2(x) + \tan^2(x)$$

using the Pythagorean identity $\sec^2(x) = 1 + \tan^2(x)$.

We can now use this identity a second time to rewrite $\sec^2(x)$, giving

$$\sec^2(x) + \tan^2(x) = 1 + \tan^2(x) + \tan^2(x) = 1 + 2\tan^2(x)$$

as required.

(e) Initially, we might think to use the difference of two squares identity on the left-hand side, as in the previous part. However, we do not have any immediate simple expression for $\sec^2(x) \pm \operatorname{cosec}^2(x)$, so this will not obviously help us.

So instead, we will try to clear the fractions, and use some or all of the identities (from the function definitions):

$$\sin(x)\sec(x) = \tan(x) \qquad \sin(x)\operatorname{cosec}(x) = 1$$
$$\cos(x)\sec(x) = 1 \qquad \cos(x)\operatorname{cosec}(x) = \cot(x).$$

We then have

$$\sec^4(x) - \operatorname{cosec}^4(x) = \frac{\sin^2(x) - \cos^2(x)}{\cos^4(x)\sin^4(x)}$$
$$\iff \quad (\sec^4(x) - \operatorname{cosec}^4(x))\cos^4(x)\sin^4(x) = \sin^2(x) - \cos^2(x)$$
$$\iff \quad \sin^4(x) - \cos^4(x) = \sin^2(x) - \cos^2(x)$$
$$\iff \quad (\sin^2(x) - \cos^2(x))(\sin^2(x) + \cos^2(x)) = \sin^2(x) - \cos^2(x)$$
$$\iff \quad \sin^2(x) - \cos^2(x) = \sin^2(x) - \cos^2(x),$$

where on the final line we have made use of the Pythagorean identity. This is clearly an identity (as both sides are identical), so we have established the original identity (at least for all values of x where both sides are defined).

(f) With reciprocal trigonometric functions on the denominator of a fraction, it seems sensible to multiply both sides to clear the fractions. We then have

$$\frac{1}{\cosec(x) + \cot(x)} = \frac{1 - \cos(x)}{\sin(x)}$$

$$\Longleftrightarrow \quad \sin(x) = (1 - \cos(x))(\cosec(x) + \cot(x))$$

$$\Longleftrightarrow \quad \sin(x) = \cosec(x) + \cot(x) - \cot(x) - \cos(x)\cot(x)$$

$$\Longleftrightarrow \quad \sin(x) = \cosec(x) - \cos(x)\cot(x)$$

$$\Longleftrightarrow \quad \sin(x) = \frac{1}{\sin(x)} - \frac{\cos^2(x)}{\sin(x)}$$

$$\Longleftrightarrow \quad \sin(x) = \frac{1 - \cos^2(x)}{\sin(x)}$$

$$\Longleftrightarrow \quad \sin(x) = \frac{\sin^2(x)}{\sin(x)}$$

$$\Longleftrightarrow \quad \sin(x) = \sin(x).$$

This is clearly true, and so our original identity is also true.

(g) We could use the same method for this identity as in the previous part. It will work, though it looks quite heavy algebraically. Another way we can proceed is to observe a relationship between the two denominators: if we write the left-hand denominator in terms of sine and cosine, we obtain

$$\sec(x) + \cosec(x) = \frac{1}{\cos(x)} + \frac{1}{\sin(x)} = \frac{\sin(x) + \cos(x)}{\sin(x)\cos(x)}.$$

We therefore see that multiplying the left-hand denominator by $\sin(x)\cos(x)$ gives the right-hand denominator, so we can proceed as follows, noting that $\tan(x)\cos(x) = \sin(x)$ and $\cot(x)\sin(x) = \cos(x)$:

$$\frac{\tan(x) + \cot(x)}{\sec(x) + \cosec(x)} = \frac{\tan(x) + \cot(x)}{\sec(x) + \cosec(x)} \times \frac{\sin(x)\cos(x)}{\sin(x)\cos(x)}$$

$$= \frac{\sin(x)\sin(x) + \cos(x)\cos(x)}{\sin(x) + \cos(x)}$$

$$= \frac{1}{\sin(x) + \cos(x)},$$

as we want.

Q8.

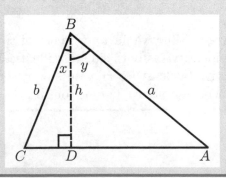

(a) Using the trigonometric formula for the area of a triangle,

$$\text{Area } ABC = \text{Area } ABD + \text{Area } DBC$$

$$\Rightarrow \quad \frac{1}{2}ab\sin(x+y) = \frac{1}{2}ah\sin(y) + \frac{1}{2}bh\sin(x)$$

Considering triangle ABD, $h = a\cos(y)$ and considering triangle DBC, $h = b\cos(x)$. Substituting these into the above,

$$\frac{1}{2}ab\sin(x+y) = \frac{1}{2}ah\sin(y) + \frac{1}{2}bh\sin(x)$$

$$\Rightarrow \quad \frac{1}{2}ab\sin(x+y) = \frac{1}{2}ab\cos(x)\sin(y) + \frac{1}{2}ba\cos(y)\sin(x)$$

$$\Rightarrow \quad \sin(x+y) = \cos(x)\sin(y) + \cos(y)\sin(x)$$

$$= \sin(x)\cos(y) + \cos(x)\sin(y).$$

(b) Replacing y by $-y$ we have,

$$\sin(x-y) = \sin(x)\cos(-y) + \cos(x)\sin(-y)$$

$$= \sin(x)\cos(y) - \cos(x)\sin(y)$$

Worked Solutions — Exercise 10.1

(a) By the definition of $|\cdot|$, we have

$$|x_i - x| = \begin{cases} x - x_i, & x \geq x_i, \\ x_i - x, & x < x_i. \end{cases}$$

Firstly, for some j, let us split $g(x)$ as follows:

$$g(x) = \sum_{i=1}^{j-1} |x_i - x| + \sum_{i=j}^{n} |x_i - x|.$$

Now suppose $x_{j-1} \leq x \leq x_j$, then we can use the definition of $|\cdot|$ to find

$$g(x) = \sum_{i=1}^{j-1} (x - x_i) + \sum_{i=j}^{n} (x_i - x)$$

$$= (j-1)x - \sum_{i=1}^{j-1} x_i - (n-j+1)x + \sum_{i=j}^{n} x_i$$

$$(2(j-1) - n)x + \underbrace{\sum_{i=j}^{n} x_i - \sum_{i=1}^{j-1} x_i}_{= \text{ constant}}.$$

Hence, for $x_{j-1} \leq x \leq x_j$, $g(x)$ is linear with gradient $2(j-1) - n$.

(b) Suppose that n is odd, then

$$j \leq (n+1)/2$$
$$\Rightarrow \quad 2(j-1) - n \leq (n+1) - 2 - n = -1,$$

hence the function is decreasing.
Similarly,

$$j \geq (n+1)/2 + 1$$
$$\Rightarrow \quad 2(j-1) - n \geq (n+1) + 2 - 2 - n = 1,$$

hence the function is increasing.
We conclude that $g(x)$ is a piecewise linear function, which must have its minimum when $x = x_{(n+1)/2}$, which is the median of the dataset.

(c) Suppose now that n is even. We consider three cases $j < n/2 + 1$, $j = n/2 + 1$ and $j > n/2 + 1$:

$$j < n/2 + 1$$
$$\Rightarrow \quad 2(j-1) - n < 0,$$

hence the function is decreasing.

$$j = n/2 + 1$$
$$\Rightarrow \quad 2(j-1) - n = 0,$$

hence the function is constant.

$$j > n/2 + 1$$
$$\Rightarrow \quad 2(j-1) - n > 0,$$

hence the function is increasing.

Thus, we have a piecewise linear function which decreases, then is constant, then increases. It must have a minimum where it is constant, so the values which minimise $g(x)$ are $x_{\frac{n}{2}} \leq x \leq x_{\frac{n}{2}+1}$. The median in this case is $\frac{x_n + x_{n+1}}{2}$, which lies in the interval $[x_{\frac{n}{2}}, x_{\frac{n}{2}+1}]$. Therefore, the median minimises $g(x)$ in this case as well.

Worked Solutions — Exercise 10.2

Since $y_i = (X_i - a)/b$ where $b > 0$, we have $x_i = by_i + a$.

$$\mu_x = \frac{1}{N} \sum_{i=1}^{N} x_i = \frac{1}{N} \sum_{i=1}^{N} (by_i + a) = \frac{b}{N} \left(\sum_{i=1}^{N} y_i \right) + a = b\mu_y + a.$$

$$\sigma_x^2 = \frac{1}{N} \sum_{i=1}^{N} (x_i - \mu_x)^2 = \frac{1}{N} \sum_{i=1}^{N} ((by_i + a) - (b\mu_y + a))^2$$

$$= \frac{1}{N} \sum_{i=1}^{N} (by_i - b\mu_y)^2$$

$$= \frac{b^2}{N} \sum_{i=1}^{N} (y_i - \mu_y)^2 = b^2 \sigma_y^2,$$

$$\Rightarrow \sigma_x = b\sigma_y.$$

Worked Solutions — Exercise 10.3

(a) We use the definition of S^2 and the linearity properties of expectation to take the expectation operator inside the sum:

$$\mathbb{E}(S^2) = \mathbb{E}\left(\frac{1}{n-1} \sum_{i=1}^{n} (X_i - \bar{X})^2 \right)$$

$$= \frac{1}{n-1} \sum_{i=1}^{n} \mathbb{E}\left((X_i - \bar{X})^2 \right).$$

(b) We recall $\bar{X} = \frac{1}{n} \sum_{j=1}^{n} X_j$, where index j has been used because i already appears in the definition of S^2. Substituting this into the result from part (a) and expanding the brackets gives

$$\mathbb{E}(S^2) = \frac{1}{n-1} \sum_{i=1}^{n} \mathbb{E}\left(\left(X_i - \frac{1}{n}\sum_{j=1}^{n} X_j\right)^2\right)$$

$$= \frac{1}{n-1} \sum_{i=1}^{n} \mathbb{E}\left(X_i^2 - \frac{2}{n}X_i\sum_{j=1}^{n} X_j + \frac{1}{n^2}\left(\sum_{j=1}^{n} X_j\right)^2\right).$$

(c) We first separate the sum into a term involving X_i and the rest, to obtain:

$$\frac{2}{n}X_i\sum_{j=1}^{n} X_j = \frac{2}{n}X_i\left(X_i + \sum_{\substack{j\neq i \\ j=1}}^{n} X_j\right)$$

$$= \frac{2}{n}\left(X_i^2 + \sum_{\substack{j\neq i \\ j=1}}^{n} X_iX_j\right).$$

We then take expectations and use the linearity of $\mathbb{E}(\cdot)$ and the result that X_i and X_j are independent if $i \neq j$:

$$\mathbb{E}\left(\frac{2}{n}X_i\sum_{j=1}^{n} X_j\right) = \mathbb{E}\left(\frac{2}{n}\left(X_i^2 + \sum_{\substack{j\neq i \\ j=1}}^{n} X_iX_j\right)\right)$$

$$= \frac{2}{n}\left(\mathbb{E}(X_i^2) + \sum_{\substack{j\neq i \\ j=1}}^{n} \mathbb{E}\left(X_iX_j\right)\right)$$

$$= \frac{2}{n}\left(\mathbb{E}(X_i^2) + \sum_{\substack{j\neq i \\ j=1}}^{n} \mathbb{E}(X_i)\mathbb{E}(X_j)\right).$$

Now, using $X \sim X_i$ (and $X \sim X_j$), we have

$$\mathbb{E}\left(\frac{2}{n}X_i\sum_{j=1}^{n}X_j\right) = \frac{2}{n}\left(\mathbb{E}(X^2) + \sum_{\substack{j\neq i\\j=1}}^{n}\mathbb{E}(X)\mathbb{E}(X)\right)$$

$$= \frac{2}{n}\left(\mathbb{E}(X^2) + \sum_{\substack{j\neq i\\j=1}}^{n}\mathbb{E}(X)^2\right)$$

$$= \frac{2}{n}\left(\mathbb{E}(X^2) + (\mathbb{E}(X))^2\sum_{\substack{j\neq i\\j=1}}^{n}1\right)$$

$$= \frac{2}{n}\left(\mathbb{E}(X^2) + (\mathbb{E}(X))^2(n-1)\right)$$

$$= \frac{2}{n}\mathbb{E}(X^2) + \frac{2(n-1)}{n}(\mathbb{E}(X))^2.$$

(d) We expand the brackets and remark that, because we have a square, we obtain a double sum. We then pull the squared terms out of the sum to obtain the result

$$\frac{1}{n^2}\left(\sum_{j=1}^{n}X_j\right)^2 = \frac{1}{n^2}\left(\sum_{l=1}^{n}\sum_{j=1}^{n}X_lX_j\right)$$

$$= \frac{1}{n^2}\left(\sum_{l=1}^{n}(X_l)^2 + \sum_{l=1}^{n}\sum_{\substack{j\neq l\\j=1}}^{n}X_lX_j\right).$$

Now we use the linearity of expectation, together with independence of X_l and X_j for $l \neq j$:

$$\mathbb{E}\left(\frac{1}{n^2}\left(\sum_{j=1}^{n}X_j\right)^2\right) = \frac{1}{n^2}\left(\sum_{l=1}^{n}\mathbb{E}(X_l^2) + \sum_{l=1}^{n}\sum_{\substack{j\neq l\\j=1}}^{n}\mathbb{E}(X_l(X_j)\right)$$

$$= \frac{1}{n^2}\left(\sum_{l=1}^{n}\mathbb{E}(X_l^2) + \sum_{l=1}^{n}\sum_{\substack{j\neq l\\j=1}}^{n}\mathbb{E}(X_l)\mathbb{E}(X_j)\right).$$

Finally, we use $X_l \sim X$ and $X_j \sim X$ to give

$$\mathbb{E}\left(\frac{1}{n^2}\left(\sum_{j=1}^{n}X_j\right)^2\right) = \frac{1}{n^2}\left(\sum_{l=1}^{n}\mathbb{E}(X^2) + \sum_{l=1}^{n}\sum_{\substack{j\neq l\\j=1}}^{n}\mathbb{E}(X)\mathbb{E}(X)\right)$$

$$= \frac{1}{n^2}\left(\sum_{l=1}^{n}\mathbb{E}(X^2) + \sum_{l=1}^{n}\sum_{\substack{j\neq l\\j=1}}^{n}(\mathbb{E}(X))^2\right)$$

$$= \frac{1}{n^2}\left(\mathbb{E}(X^2)\sum_{l=1}^{n}1 + (\mathbb{E}(X))^2\sum_{l=1}^{n}\sum_{\substack{j\neq i\\i=1}}^{n}1\right)$$

$$= \frac{1}{n^2}\left(n\mathbb{E}(X^2) + n(n-1)(\mathbb{E}(X))^2\right)$$

$$= \frac{1}{n}\mathbb{E}(X^2) + \frac{n-1}{n}(\mathbb{E}(X))^2.$$

(e) Finally, we see that, from part (b) and using the results from (c) and (d),

$$\mathbb{E}\left((X_i - \bar{X})^2\right) = \mathbb{E}\left(X_i^2 - \frac{2}{n}X_i\sum_{j=1}^{n}X_j + \frac{1}{n^2}\left(\sum_{j=1}^{n}X_j\right)^2\right)$$

$$= \mathbb{E}(X_i^2) - \mathbb{E}\left(\frac{2}{n}X_i\sum_{j=1}^{n}X_j\right) + \mathbb{E}\left(\frac{1}{n^2}\left(\sum_{j=1}^{n}X_j\right)^2\right)$$

$$= \mathbb{E}(X^2) - \frac{2}{n}\mathbb{E}(X^2) - \frac{2(n-1)}{n}(\mathbb{E}(X))^2 + \frac{1}{n}\mathbb{E}(X^2)$$

$$\quad + \frac{n-1}{n}(\mathbb{E}(X))^2$$

$$= \left(1 - \frac{2}{n} + \frac{1}{n}\right)\mathbb{E}(X^2) + \left(-\frac{2(n-1)}{n} + \frac{n-1}{n}\right)(\mathbb{E}(X))^2$$

$$= \frac{n-1}{n}\left(\mathbb{E}(X^2) - (\mathbb{E}(X))^2\right)$$

Then

$$\mathbb{E}(S^2) = \frac{1}{n-1}\sum_{i=1}^{n}\frac{n-1}{n}\left(\mathbb{E}(X^2) - (\mathbb{E}(X))^2\right)$$

$$= \mathbb{E}(X^2) - (\mathbb{E}(X))^2$$

$$= \text{Var}(X).$$

CPSIA information can be obtained
at www.ICGtesting.com
Printed in the USA
LVHW060730150623
749582LV00002B/4